HARVARD UNIVERSITY.

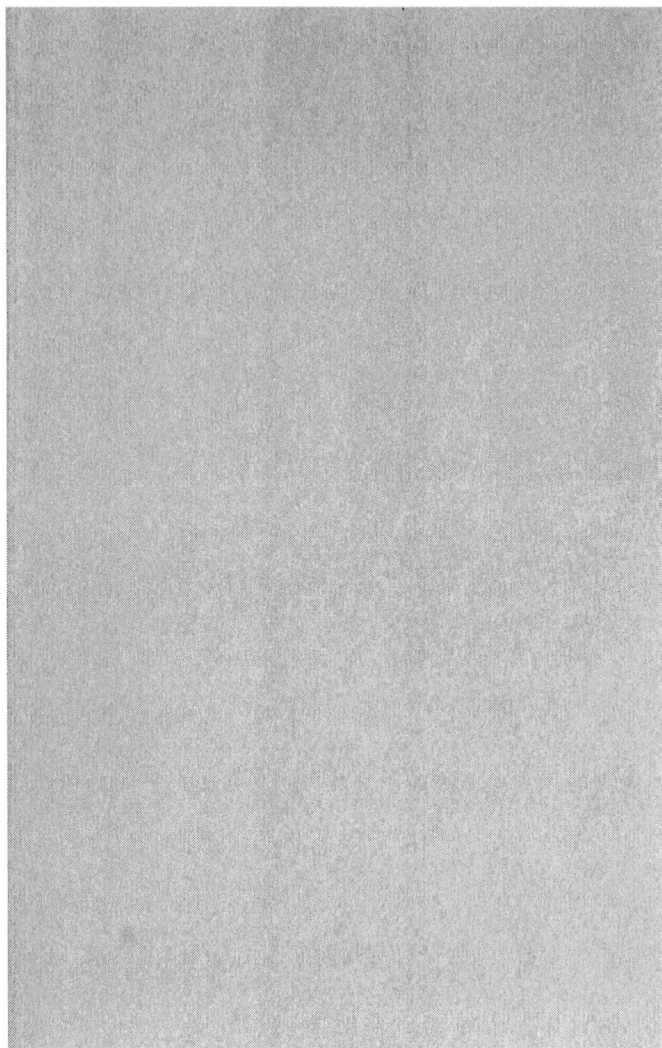

A

TREATISE

ON

BRITISH SONG-BIRDS.

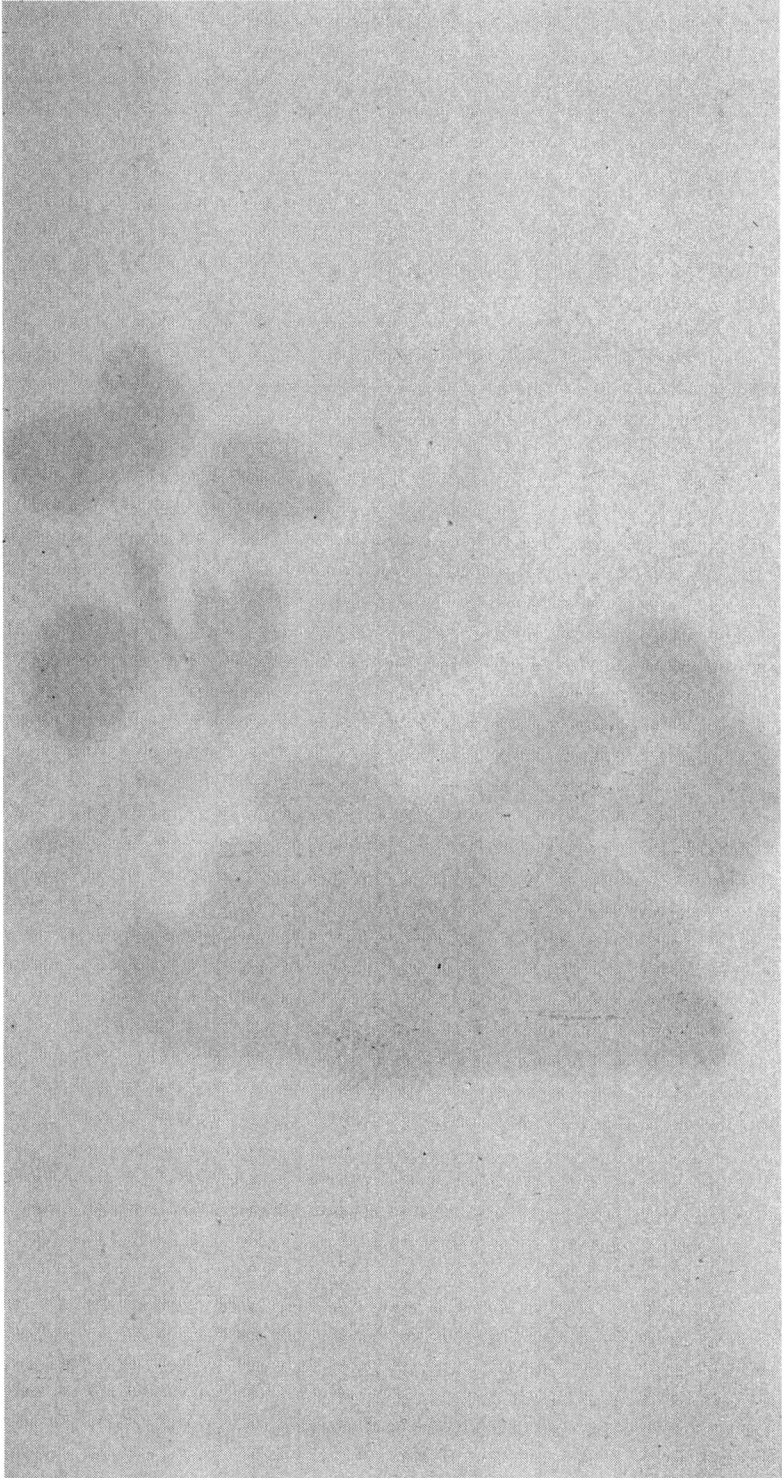

R. Scott Sculp.ᵗ

NIGHTINGALE.

Published by John Anderson Junᵣ 55. North Bridge Street Edinburgh. 1823.

A

TREATISE ON BRITISH SONG-BIRDS.

INCLUDING

OBSERVATIONS ON THEIR NATURAL HABITS, MAN-
NER OF INCUBATION, &c. WITH REMARKS ON
THE TREATMENT OF THE YOUNG AND
MANAGEMENT OF THE OLD BIRDS
IN A DOMESTIC STATE. ⌐ by Patrick Syme ⌐

WITH FIFTEEN ENGRAVINGS.

JOHN ANDERSON, JUN. EDINBURGH,

55, NORTH BRIDGE-STREET;

AND SIMPKIN & MARSHALL, LONDON.

———

MDCCCXXIII.

4488
5.

INTRODUCTION.

―――――――

THOUGH men congregate in cities, it may be pre-
sumed, from various circumstances, that they assemble there more from necessity, than any preference founded on natural love for a town-life.
Men do not naturally prefer crowded streets, confined alleys, or dusty rooms, to woodland walks,
grassy paths, or green arbours. Do not those,
whose professions prevent them from getting to
the country, endeavour to bring it as much as possible to them? What are town-gardens, and shrubberies in squares, but an attempt to ruralize the
city? So strong is the desire in man to participate

A

in country pleasures, that he tries to bring some of them even to his room. Plants and birds are sought after with avidity, and cherished with delight. With flowers, he endeavours to make his apartments resemble a garden;—and thinks of groves and fields, as he listens to the wild, sweet melody of his little captives. Those who keep and take an interest in song-birds, are often at a loss how to treat their little warblers during illness, or to prepare the proper food best suited to their various constitutions. But this knowledge is absolutely necessary to preserve these little creatures in health: For want of it, young amateurs and bird-fanciers have often seen, with regret, many of their favourite birds perish.

Several detached little works on this subject have already been given to the public: but these contain little more than rules for rearing and managing a species, as the Nightingale or Canary, or, at most, a genus, with slight remarks on other song-birds. It therefore appeared to the publisher of the present volume, that a more comprehensive work on the subject was much wanted. For these reasons the present Treatise was undertaken, and we trust it will be found to contain much useful information on the subject of song-birds.

In arranging this work, we have availed ourselves of the knowledge of our friends, to which we have joined our own practical experience: and, to render the whole more complete, we have given a description of each bird, together with observations on their habits in a wild state, with some remarks and anecdotes to illustrate their manners in a state of domestication.

The delightful music of song-birds is, perhaps, the chief cause why these charming little creatures are, in all countries, so highly prized.—Music is an universal language;—it is understood and cherished in every country: The savage, the barbarian, and the civilized individual, are all passionately fond of music, particularly of melody. But, delightful as music is, perhaps there is another reason that may have led man to deprive the warblers of the woods and fields, of liberty, particularly in civilized states, where the intellect is more refined, and, consequently, the feelings more adapted to receive tender impressions;—we mean the associations of ideas. Their sweet melody brings him more particularly in contact with groves and meadows—with romantic banks—or beautiful sequestered glades—the cherished scenes, perhaps, of his early youth.

But, independent of this, the warble of a sweet
song-bird is, in itself, very delightful;—and, to
men of sedentary habits, confined to cities by pro-
fessional duties, and to their desks most part of
the day, we do not know a more innocent or more
agreeable recreation than the rearing and training
of these little feathered musicians.

Though the British song-birds belong to six
different genera, viz. the Stare, Thrush, Lark,
Warbler, Finch, and Gross-beak, Willoughby has
very properly (at least in our opinion,) divided
them into two divisions, and distinguished them
by the terms, Hard and Soft-billed Birds. These
two divisions are well marked. The birds of the
first division differ considerably from those of the
second in their habits, in their food, and in the
melody of their songs. The soft-billed birds, in a
wild state, are shy and timid. With a few excep-
tions, they keep at a distance from man. They
construct their nests in retired situations, and
conceal them with great ingenuity. Their food,
in summer, consists of insects, viz. earth-worms,
snails, slugs, spiders, earwigs, sklaters, or hog-lice,
flies, smooth caterpillars, ants, ant-eggs, centipedes,
&c.; and in winter they eat wild berries, such as
the berries of ivy, misletoe, privet, holly, haw-

thorn, barberries, &c. Their song is a rich wild
cadence, of sweetly-varied mellow, plaintive notes.
The young are difficult to rear from the nest;
and the constitutions of both young and old birds
are rather delicate. They are grateful and affectionate little creatures. Some have even been
known to pine and die, when deprived of those
they were attached to. None of the soft-billed
birds have ever been known to breed in a state of
captivity. At the head of this tribe may be placed
the nightingale, followed by the blackcap, woodlark, redbreast, pettychaps, thrush, skylark, &c.
The hard-billed species, with the exception of the
redpoles and linnets, draw near the habitation of
man, and nestle in his neighbourhood. They feed,
in spring, on the young buds of trees; and, in summer and autumn, on grain of different kinds, and
the seeds of hemp, rape, lint, plantain, chickweed,
groundsel, thistle, and other plants; and, in winter,
on wild berries, grain, &c. Their song is cheerful, varied, loud and powerful, often shrill and
piercing, though some of them have a low, sweet
warble; but it is seldom mellow or plaintive.
Their constitutions are generally hardy, and the
young are easily reared from the nest. Some of
them breed in captivity, as the siskin; and also
several of them with the canary, as the goldfinch,

linnet, yellowhammer, &c. Like the soft-billed birds, they get much attached to those who are kind to them; and some of them have been known to die when deprived of their benefactors. They are also susceptible of friendship amongst themselves. Foremost in this division ranks the canary, followed by the siskin, linnet, redpole, goldfinch, bullfinch, &c.

It is rather remarkable that all our finest songbirds have but few showy colours in their plumage. This, we believe, has been observed with respect to the best songsters of both the old and new continents; while those possessed of beauty and brilliancy of plumage have generally neither melody nor song;—no notes, but what are disagreeable screams, or harsh grating sounds. How finely this illustrates the difference between modest merit, dressed in a plain garb, and that gaudy, meretricious, external show, which vain and silly individuals think will supply the place of internal worth.

Ought not the operations of Nature, or, in other words, the power and benificence of that great and good Being, who directs all her functions, and whose divine influence formed and pervades all created matter, continually to call forth our admiration?—and ought not our hearts

to overflow with gratitude to Him, who has so
graciously conferred upon us so many blessings,
and, amongst them, the capability of deriving pure
and innocent pleasure from the investigation of
his works? Is not a mite as great a wonder as
an elephant?—a tuft of moss as that of an oak-
tree?—and the process employed in forming the
feather of a wren, does it not display as much
power and wisdom as the formation of an eagle?
We cannot look at the plumage of a bird, and re-
flect on the wonderful arrangement of parts necessary for the production of the feathers, without
being lost in astonishment! Feathers seem to us to
have both secreting vessels for their nourishment,
and vessels for secreting the matter which gives the
feathers their colour, and, when the feathers are
completed, the colouring vessels lie inactive till
the next moulting. But these vessels may be
brought partially into action, to renew feathers that
have been lost by accident, or to renew the tail of
a bird that has been pulled for the pip; it being
considered sometimes beneficial, during that disease, to draw the feathers from a bird's tail. Feathers we conceive to be coloured much in the same
manner as flowers. Each colour has a set of vessels peculiar to itself, and totally independent of

the blood-vessels that nourish the feathers. For instance, if a feather is all one tint, it will have only one secreting gland, and one vessel running along the quill, which is afterwards branched out and ramified all over the feather through which the colouring matter is injected; but, if the feather has three colours, as red, blue, and yellow; then there will be three distinct glands, and one vessel attached to each gland, for conveying the colouring matter to the feather, and so on in proportion to the number of tints in each feather.

The annual loss and renewal of feathers, at the period of moulting, is a curious circumstance in the natural history of birds; but not more singular than the shedding and renewing of the horns of a deer, or the casting and reproducing of the shell of a crab. Animals that undergo the processes of the loss and reproduction of parts, are, during that operation, more or less affected with temporary debility and sickness. In a wild state, nature, it is supposed, carries birds through the operation without injury; but, in a state of domestication,—their natural habits and constitutions having undergone a change,—they require both care and attention at that time. They ought to be kept in a dry, well-aired room, the temperature of which should be about sixty degrees. In the middle of the day, if the sun shines, and the air is dry, the window may be opened for a few hours. Their diet, at this time, ought to be very nourishing, with, occasionally, a little green food; such as lettuce leaves, lettuce-seed, chickweed, cherries, a ripe soft pear or a roasted apple; also, a little maw-seed, as it tends to cheer and keep them in spirits.

The nest of a bird is a wonderful fabric when we consider the architect. Directed by instinct, it chooses materials, and arranges them in a manner best suited for the purposes of incubation,— and this according to the size of the bird. How different is the structure of the nest of a wren and that of an eagle! The nest of an eagle is formed of sticks laid in a horizontal position, upon which a few stalks of fern are laid, and on this the eggs lie, exposed to the air, when the female is absent. That of a wren is a large bunch of moss, artfully woven together, and covered all round, except a small hole near the top, and lined with soft and warm materials. Here the pea-like eggs of the wren lie, without being affected by the cold when the female is absent in search of food; for the female never sits constant, till within a few days of the brood being hatched. Sometimes, indeed, the cock-bird takes her place,—sometimes he feeds her while she is on the nest,—but female birds always sit very close for a few days prior to the exclusion of the young. The nests of the rook, magpie, jay, wood-pigeon, &c. are pervious to the air throughout, yet the eggs are duly hatched; while the eggs of the golden-crested wren are enveloped in down and feathers. Instinct seems to direct the bird to construct and suit the nest to the size of the eggs. But it appears to us, that the eggs of birds, even of the same size, require a different temperature. For instance :—the nest of the sedge-warbler is a thin flimsy structure, and placed in a damp exposed situation ; while that of the common house-sparrow is formed of moss, feathers, thread, and other warm materials, thickly lined with feathers, and placed in a warm spot. We cannot account for this in any other way than by supposing that the eggs of the house-sparrow require a higher degree of temperature to hatch them than those of the sedge-warbler. Another thing that has struck us as singular, with regard to the materials birds use in forming their nests, is, that the feathers or hair, which they make choice of for lining them, are always white or grey—never black. Whether the white colours

of the feathers, &c. have any thing to do in con-
centrating the heat, or that black might conduct
the heat through the nest by radiation, and thus
allow it to escape, we cannot say; but we can
vouch for the fact,—that the feathers and hair,
found in the insides of the nests of birds, are always white, grey, or light-coloured.

The materials employed by birds, in forming
their nests, are, perhaps, better criterions to judge
of the species by, than even the colours of the eggs,
which, in very different genera, as well as in distinct species, are often very similar in tint; while,
amongst the eggs of the same species, a considerable difference may be found with regard to the
colour of their spots or freckles. The colour of and
the freckles on eggs appear to us to be produced
much in the same manner as the tints and spots
on the shells of snails; that is,—we conceive the
colouring matter is forced through the shell by
some process unknown to us, from the inner to
the outer surface, where it appears brightest; but
with this difference,—that the colours of the eggs
of a bird bear no analogy to that of the matter under the shells, or to the feathers of the parent-
bird; whereas many testaceous animals,—such as
snails,—are found to be marked with colours si-

milar to those on the exterior surface of their
shells.

We have taken the liberty to introduce into this
treatise a few birds that are considered by some
authors as destitute of song, but which we know,
from our own experience, to possess a very sweet
warble. To these, perhaps, we should have added the lesser pettychaps, or chipchop, the greater
and lesser whitethroats, the grasshopper-war-
bler, and the large and blue titmice, or tomtits,
as some consider them song-birds; but having
only heard the cricket-like note of the grasshop-
per-lark or warbler—the churring notes of the
whitethroats—and the simple chipchip of the lesser pettychaps,—and no other warble from these
birds,—we thought it better to leave them out for
the present. The blue titmouse has a kind of
shrill song, but it is too monotonous to entitle the
bird to be ranked as a warbler. The great titmouse is a better songster: indeed it has a low
warble, which we consider rather sweet; but this
genus is so timid and wild,—at least those we
have seen in cages,—that we think they are not
adapted to a state of captivity; because, though
they eat freely, yet they seldom, if ever, become
tame. Some think the lesser whitethroat an excellent warbler, and others say, that the lesser

pettychaps, or lesser fauvette, is a mocking-bird.
Some amateurs even keep in their aviaries the
whin-chat, the stone-chat, and the white-rump;
Though well acquainted with these birds, as far
as we know, they have no song, only a single note,
two or three times repeated; we have therefore
not given them a place in this work. One of the
birds we have given as a distinct species, and an
excellent songster, (we mean the solitary thrush,)
is said to be the young of the stare or starling;
but the reasons why we think it a distinct species
are these :—it is said to have all the appearance
of a matured bird in full adult feather—its plumage is lighter, and more of a brown colour than
that of the young starling, and bird-fanciers know
it has a most excellent natural song. Now, this
last is not the case with either young or old starlings; besides, young starlings, at least those we
have had, are of a dull, dingy black, somewhat like
the young of the black and ring ouzels,—they do
not seem matured—their feathers hang loose about
them—they have a bunched appearance—and possess all the characteristics by which young birds
are generally distinguished.

We may here notice, that the linnet, grey linnet, brown linnet, red-breasted linnet, and greater

redpole, are, by Linnæus and Montagu, considered as the same bird, only differing in plumage,
according to age and season. The plumage of the
greater redpole is considered by Montagu to be the
perfect plumage of the species. He says he has
seen them in all stages, from the brilliant red on
the forehead and breast, to the faintest appearance
of the red on these parts, while others had no red
on either. He also says, young redpoles, brought
up from the nest, never assume the red plumage,
and that the adult redpoles, when confined in a
cage, lose the red, and never acquire it again. It
is with great deference we offer an opinion contrary to the authority of these great naturalists;
but we are rather inclined to think, that the linnet,
grey linnet, or brown linnet, the red-breasted or
rose linnet, and the greater redpole, are three distinct birds, but as closely allied to each other as
the three species of wagtails are to one another.
We shall state our reasons for thinking so. The
grey linnet is rather less, and more slender than
the redpole. The white on the quills and outer
feathers of the tail of the linnet is broader and
brighter than that of the redpole. The bill of the
redpole is rather larger, and broader at the base,
than that of the linnet. The marks on the breast

of the redpole, in all its changes of plumage, run
in decided streaks, pointing downwards, while
those on the breast of the linnet are much fainter,
and more inclining to an irregular mottled appearance. The eggs of the redpole are of a bluish
white colour, marked with specks and lines of
dingy purple,—those of the linnet are reddish white,
freckled with small spots of brownish orange; but,
above all, the songs of the two birds are different:
Both are good, but we think that of the linnet
the best; and the song of the red-breasted linnet
differs from both. Now, if our own experience is
correct, we have, from observation, been always
led to believe that the song of all birds of the
same species, in a wild state, is invariably the
same. It is true, the habits of these three birds
are very similar. They frequent the same places,
and build their nests in the same kind of bushes,
which are generally furze or whins; but the redpole is much more wild and shy than the linnet.

In Scotland there are vast numbers of grey linnets, but the redpole and red-breasted linnet are
rather rare birds in that country;—a thousand
grey linnets may be found for one redpole, and
five or six hundred grey for one red-breasted linnet. In spring and summer we have often had

the nests of grey linnets, and seen numbers of
them shot in winter, spring, summer, and autumn, but always found the plumage nearly the
same;—it is lighter, and more of a greyish brown,
than either the red-breasted linnet or redpole.
The plumage of the last birds is more of a tint between amber and chesnut-brown;—that of the
grey linnet is between yellowish and hair-brown;
but the aspect, as well as the colours of the three
birds, are very different from each other. The
grey linnet never has red on its breast;—the red
on the breast of the red-breasted linnet is pale,
and of a tint between carmine and lake red, softly
waved transversely;—that on the breast of the
redpole is deep artereal blood-red, streaked downwards and strongly marked. That redpoles, in a
state of confinement, lose the red altogether, may
be accounted for by change of food, or the priva-
tion of something they were accustomed to in a
wild state. Hempseed changes the plumage of
the bullfinch to black,—and very long confinement,
or age, or some other cause, affects the goldfinch,
so that it assumes a white appearance; but we never could perceive any difference in the plumage
of wild and tame linnets. For these reasons, we
are led to conclude that the grey or brown lin-

net, the red-breasted or rose linnet, and the greater redpole, are three distinct species, but, in many respects, closely allied to each other.

Some have even thought the twite, or mountain-linnet, only a variety of the grey linnet, and the lesser redpole a variety of the twite; but the habits of these birds are different, and those who examine them together will readily perceive, by the different aspects of all these birds, that they are well marked and distinct species.

In a popular treatise like the present, little will be expected regarding the habits of song-birds in a state of liberty. Observations of this nature are more properly connected with scientific works on their natural history; but we have endeavoured to illustrate, as much as possible, their manners in a state of captivity, by anecdotes, which, to us at least, render these little creatures very interesting objects of attention, independent of their sweet melody. Most of the hard-billed birds' are stationary with us the whole year. Some of them, however, migrate locally, from north to south, in winter; but the soft-billed species are generally summer birds of passage, and leave the island in autumn and return to it in spring.

By some observations made on the soft-billed

species; by Mr Swede, a very eminent amateur and bird fancier, it appears that these birds, even in a state of captivity, are affected by instinctive feelings to migrate; for that gentleman observed that the birds in his possession became unusually agitated at the periods at which birds of the same species leave the island in autumn, or return to it again in the spring. He also noticed, that they were affected in a similar manner twice during the winter; from which he concludes that these birds, after they leave us, perform two other migrations. He says, the fit of restlessness came on towards night,—that they appeared agitated, fluttered much with their wings, and went from one end of the perch to the other, like birds that wished to fly upwards, as they held their heads up, and their eyes were directed to the ceiling of the room: So fixed, indeed, was their stare, and so overpowered did they appear to be by some internal feeling, that they took no notice of any thing held near them, not even a lighted candle. Mr Swete says, that the fit lasted, in autumn and in spring, for a few hours every evening, during a fortnight,—while, in winter, the birds were affected each time only for a few evenings; from which that gentleman concludes that their winter

flights are much shorter than those of their autumnal and spring journies. All our summer
birds of passage come from the south in spring
and summer, perhaps to avoid the excessive heat
of southern climates, and return southwards in
autumn, while all our winter birds of passage
come from the north in quest of food and perhaps
to avoid the cold, and in spring retire to the north,
for the purposes of incubation, &c.; from which it
appears that all birds of passage, north of the
line, migrate northwards in summer and southwards in winter; and, from analogy, we may sup-
pose, that birds of passage south of the line, fly
southwards in summer and return north in winter, to avoid the cold, &c.

It is not easy to explain why birds, in this
island, prefer one county to that of another; or
why they haunt only certain parts of a county;
while the surrounding districts abound with covert to afford them shelter, and with food such
as they are supposed to delight in. This can only
be accounted for by supposing, that they came
there originally by accident, and that they returned, after migrating, to the same place where they
first built their nests, and that their young would
return to the spots where they themselves were

bred. At present, some species are very local in their haunts, as the nightingale, woodlark, red-poles, &c.; others, whether constant inhabitants or annual visitors, are very generally diffused throughout the island. The spreading of a species, perhaps, was occasioned by the birds getting too numerous for the districts they frequented, and the young were compelled, by older and stronger birds, to seek out haunts for themselves. Several instances might be given, to show, that birds are now found in places in this island which they formerly did not frequent;—but we shall merely mention one regarding the blackbird.

This bird, though now a very common songster in almost every part of Britain, was not so sixty years ago; for, in some counties in the north of Scotland, at that period both it and the thrush were unknown. As this is a curious circumstance in the history of birds, we shall mention a fact concerning the first appearance of the blackbird in Aberdeenshire. About the year 1760, the proprietor of a beautiful property on the river Dee, (who had spent many years of his life in England,) often remarked that his woods and shrubberies were only vocal with the notes of some of the smaller song-birds, and regretted that

the blackbird and thrush never paid them a visit.
A sister of that gentleman, who, all her life, had
resided on the banks of the Dee, and who knew
no other warblers than the lark or linnet, one Sunday related to her brother, with considerable sur-
prise, that she had, that morning, heard some one
whistle some very pretty notes; but, in addition
to this strange occurrence, (whistling in Scotland,
on the Sabbath-day, was then unknown,) there
were two singular attendant circumstances :—the
one was, that the notes were often repeated, but
always the same, and the other, that, although the
whistler seemed very near, she could not perceive
him after the strictest search. A Sabbath in Scotland, even now, is sacred to peace and rest, but,
sixty years since, it was a day of universal repose.
The gentleman, therefore, from the description of
the whistle, hurried to the thicket where the notes
were said to have been heard, and there had the
pleasure of hearing, as he had anticipated, the
blackbird's rich, mellow, and animated lay.

To preserve birds in health that are kept in captivity, we think the food they get should be the
same, or as near as possible to that which they
are known to feed on in a state of liberty. But
the soft-billed birds feed on insects, and insects

cannot be got in winter; we would therefore recommend that a stock of these should be laid in
during summer, to serve them through the winter, both for food and medicine, as occasion requires. For this purpose we would suggest that
ants' eggs be gathered, along with some of the
earth of an ant's hill, and put into a tin box or
canister, and kept in a cool place; for, if too
warm, the heat might hatch the eggs. Soft grubs,
meal-worms, and even earth-worms, may be preserved in a similar manner. Mr Swete has given
a recipe to preserve flies, which is,—to gather a
quantity of the common house-flies and the large
blue-bottle or flesh-flies, and put them into a
box, and keep them dry. When they are to be
given to the birds, nothing more is necessary than
to moisten the flies with lukewarm water, and
the birds will take them freely. Spiders, earwigs,
sklaters, centipedes, and several other insects, we
think might be preserved in the same manner;
also, wild berries, of different kinds, might be
stored up for the soft-billed birds; and they should
have, occasionally, fresh damp earth, as well as
gravel stones, on the bottom of their cages: and
larks, every two or three days, ought to have a
fresh turf given to them, or they will not thrive.

Even hard-billed birds, when unwell, ought to get an insect offered to them now and then : for, though it is said by many, that these birds feed on grain and seeds only, we have seen sparrows, chaffinches, greenfinches, goldfinches, &c. pursue and seize flies of different kinds,—perhaps it was for their young,—it being our opinion that the food of the young of all song-birds is insects ; but still we think insects ought occasionally to be offered to adult birds.

Small birds kept in confinement are sometimes affected with low spirits, and more particularly during the moulting season. They then seem to loathe their food. During these fits of dullness, their diet ought to be changed, or, what perhaps is better, different kinds of food ought to be offered, and instinct will direct them to choose what is best suited to their different constitutions ; but a store of food ought to be kept for the purpose— insects, berries, fruits, &c. for the soft-billed birds ; and for the hard-billed, different kinds of seeds, such as plantain, chickweed, groundsel, lettuce, thistle, canary, rape, maw, and hemp seeds. The last ought to be given very sparingly, only a few grains at a time ; for too much hempseed, we think, is deleterious to all birds : it certainly produces a rank state of body, which is apt to change the colour of their plumage, by affecting the vessels which secrete the fluid that gives the colour to their feathers. And, if an over-dose be given to birds that have not been at all previously accustomed to it, it not unfrequently occasions death in the course of a few hours. We know, that, in the course of last summer, a fine bullfinch and canary were lost in this manner.

It may be thought a trivial circumstance to notice the perches or sticks that birds roost upon, but we know it to be of some importance; for if the sticks are too small in diameter, it tends to produce cramp in their feet; the perch, therefore, ought to be proportioned to the size of the bird, and well rounded and smooth, that it may be easily scraped and kept clean, otherwise their feet would get clogged, which injures their health. The perch for a thrush, blackbird, or starling, ought to be rather more than half an inch in diameter; and, for birds of the size of a nightingale, or goldfinch, about a quarter of an inch in diameter; and for wrens, &c. rather less. Sky-larks need none, as they never perch.

The pure attachment of birds to each other, during the important business of incubation, and their care of, and attention to their young, may be held up as examples of parental conduct even to man himself. They continue on unwearied wing, from morning to night, providing for their helpless offspring—they distribute impartially the food thus laboriously procured—they shew the most tender anxiety when danger approaches their brood—and they fearlessly expose themselves to an enemy to ward off the threatened injury. Do not these actions merit our admiration? Does not the parental conduct of these little creatures display an instinct that seems very nearly allied to moral feeling? And may not man follow their example with honour to himself and benefit to his species? Does not his own peace of mind depend much on his children's happiness? And does not parental kindness or neglect tend to increase or diminish the mass of human misery?

The length of time an egg remains in the nest, before it is hatched, depends on the size of the bird and egg. The eggs of the thrush, blackbird, and stare, require from seventeen to eighteen days; —those of the skylark from fifteen to sixteen;— and smaller birds from thirteen to fifteen, before the young appear. The operation varies about

B

twenty-four hours, according to the temperature of the atmosphere. Heat accelerates—cold retards it. The young remain in the nest till they are able to fly a little, but not sufficiently to protect them from danger. At that time the parents may be seen leading them into covert; and, if an enemy appears, the old birds instantly give the alarm, which the young perfectly understand, and they keep quiet. How old birds know their own young, and young birds their own parents, amongst so many of the same species, is, and perhaps ever will remain, a mystery to us. Birds continue to feed their brood after they have left their nest, till the young are able to feed themselves, and which they generally can do in about three weeks or a month;—then the tie between the parent and the young, in some species, is broken. Others continue in small flocks of parents and their brood, till they migrate and leave the island together. Some species, that remain with us all the year, also keep in brood-flocks through the winter, as the stare; while others, as the linnet, redpoles, &c. congregate in great numbers, and form large flocks, each flock being formed of many brood-flocks joined together; and

they remain so, during the winter, till they separate in spring, for the important purposes of incubation.

The song of birds has occasioned considerable
difference of opinion amongst naturalists; but we
believe it is now decided to be the language of
love and that of defiance : and in this opinion we
agree. Birds differ much with regard to the excellence of their song, (by which we here mean
their warble or musical notes,) both as to its
sweetness and its duration. The notes of soft-billed birds are finely-toned, mellow, and plaintive;
—those of the hard-billed species are sprightly,
cheerful, and rapid. This difference proceeds
from the construction of the larynx, as a large
pipe of an organ produces a deeper and more mel-
low-toned note than a small pipe; so the trachea
of the nightingale, which is wider than that of the
canary, sends forth a deeper and more mellow-
toned note. Soft-billed birds, also, sing more from
the lower part of the throat than the hard-billed
species. This, together with the greater width of
the larynx of the nightingale and other soft-billed
warblers, fully accounts for their soft, round, mellow notes, compared with the shrill, sharp, and clear
notes of the canary and other hard-billed songsters.

.. In a comprehensive sense, the complete song of
birds includes all the notes they are capable of ùt-
tering; and, taken in this sense, it is analogous to
the speech of man. It is the vehicle through
which these little creatures communicate and convey to each other their mutual wishes and their
wants. It may be divided into six distinct separate sounds or parts, each of which is very expressive, even to us, of the feelings which agitate
the bird at the moment. To describe their song
more fully, we shall divide it in the following
manner :—*First,* The call-note of the male in
spring; *second,* The loud, clear, ardent, fierce
notes of defiance; *third,* The soft, tender, full,
melodious, love warble; *fourth,* The notes of fear
or alarm when danger approaches the nest; *fifth,*
The note of alarm or war-cry when a bird of prey
appears; *sixth,* The note the parent birds utter to
their brood, and the chirp or note of the young.
The note of the young may be again divided into
two,—that which they utter while in the nest, and
the chirp after they have left it,—for they are very
distinct sounds or notes; to which may be added,
a soft, murmuring kind of note, emitted by the
male while he is feeding the female in the nest,
and also by her while she is receiving the food.

The call-note—the warble of love—and the notes of defiance, or prelude to battle—seem only to be understood by birds of the same species, at least in a wild state. Perhaps in a state of domestication, birds of different genera, if nearly allied, may partially comprehend these notes, as the canary bird does the notes of the siskin, the goldfinch, and the linnet. But this, we think, is more occasioned by necessity than choice in these birds; and, in this case, it is man who breaks down the barriers Nature has so wisely put between different species.

The note of fear or alarm of the cock-bird, by which he gives notice to the hen of the approach of danger near the nest, and which she perfectly understands,—for she either keeps close, or quietly makes her escape; this note we think is also only comprehended by birds of the same species, though we have certainly seen birds of different genera appear as if alarmed by this note of fear sounded by a bird of a different species or genus; but whether it was the note that alarmed them, or our presence, we cannot say. But, we are pretty sure, the notes of parent birds, and the chirp of their young, are only understood by birds of the same species, or rather we should say family for it appears to be a family language, understood reci-

procally by parent birds and their young; for the
young know the notes of their parents, and the
parents those of their own brood, amongst all the
young broods of other birds of the same species
in the neighbourhood; and this they do, as distinctly as the ewe knows the bleat of its own
lamb, or the lamb the cry of its own mother,
amongst a large flock. With regard to the note
of alarm, birds send forth on the approach of their
natural enemies, whether a hawk, an owl, or a
cat, we consider it to be a general language perfectly understood by all small birds, though each
species has a note peculiar to itself. This note
differs in sound from the note of fear or alarm
given by them when man approaches near their
nests.—This last seems confined to a species,—
but this general alarm note, (which is understood by all small birds,) we would call their war-
whoop or gathering-cry,—for it is a true, natural
slogan. All the notes comprised in the song of
birds convey delight to the mind of a lover of nature; but the bird-fanciers only prize their love-
warble and notes of defiance;—these notes, and
these only, he considers to be their song. The
musical notes of birds, whether of love or war, are
sweet, and really charming in themselves; but

they perhaps pour on the mind a greater degree of
pleasure than mere sound is capable of conveying,
—we mean the recollections of youthful days—of
endearing incidents—or of scenes connected with
country pleasure. We ourselves prefer the mellow, plaintive melody of the soft-billed species;
but others give the palm to the cheerful warble of
the hard-billed tribe : which of these two styles is
the sweetest melody we cannot determine. Both
warbles may be equally fine ; and the preference,
perhaps, may depend on taste and feeling. But
it is allowed by all who have an ear for music,—
or rather, we should say, who have an ear and
love for simple natural melody,—that the song or
warble of birds is truly delightful; but all their
musical notes cease as soon as the brood is hatched.

Before quitting the natural song, or musical
notes of British song-birds, we may here mention,
that the Honourable Daines Barrington has given
a table or scale of the comparative excellence of
the natural musical notes of British birds, exem-
plified by numbers, twenty being the point of perfection. It is as follows :—

	Mellowness of Tone.		Spright-liness.		Plaintive-ness.		Com-pass.		Execu-tion.
Nightingale	19	-	14	-	19	-	19	-	19
Skylark	4	-	19	-	4	-	18	-	19
Woodlark	18	-	4	-	17	-	12	-	8
Titlark	12	-	12	-	12	-	12	-	12
Linnet	12	-	16	-	12	-	16	-	18
Goldfinch	4	-	19	-	4	-	12	-	12
Chaffinch	4	-	12	-	4	-	8	-	8
Greenfinch	4	-	4	-	4	-	4	-	6
Hedge Sparrow	6	-	0	-	6	-	4	-	4
Aberdevine	2	-	4	-	0	-	4	-	4
Redpole	0	-	4	-	0	-	4	-	4
Thrush	4	-	4	-	4	-	4	-	4
Blackbird	4	-	4	-	0	-	2	-	2
Robin	6	-	16	-	12	-	12	-	12
Wren	0	-	12	-	0	-	4	-	4
Reed Sparrow	0	-	4	-	0	-	2	-	2
Blackcap, or Norfolk Mock Nightingale	14	-	12	-	12	-	14	-	14

From this table, and this gentleman's opinion, we beg leave to differ, as we have heard the notes of all the birds he has mentioned; and this difference of opinion either arises from our not fully comprehending the terms he has used, or from a dissimilarity of taste and feeling.

We are aware that amateurs of song-birds may have favourite warblers, or may prize the song of one bird above that of another; but we cannot understand why the honourable author, in appearance at least, seems to have confounded and misapplied terms, which to us appear so perspicuous and so well known. To shew how far, and where our opinion differs from that of the honourable author's, we think the best way is to give a table of our own, which includes some excellent British song-birds overlooked by that author in his table, —twenty still being considered the point of perfection, in our table as well as his. It is as follows:—

	Mellowness.	Sprightliness.	Plaintiveness.	Compass.	Execution.	Duration.
Nightingale	19	10	19	19	19	19
Blackcap	19	6	16	10	10	8
Redbreast	16	6	16	8	8	8
Greater Pettychaps	10	12	14	14	14	12
Redstart	6	8	6	12	10	6
Willow Wren	8	8	10	12	10	7
Golden-crested Wren	2	6	2	8	6	5
Common Wren	1	8	1	4	4	4
Sedge Warbler	0	8	3	4	4	5
Hedge Warbler	0	4	2	2	3	3
Blackbird	16	4	8	8	6	7
Thrush	16	6	10	12	8	9
Solitary Thrush	18	4	19	12	10	10
Missel Thrush	18	4	10	10	8	10
Skylark	2	19	4	18	19	18
Woodlark	14	10	16	14	14	14
Fieldlark	2	10	2	10	12	8
Titlark	4	8	4	8	6	6
Siskin or Aberdevine	0	16	2	10	12	8
Goldfinch	0	16	2	8	12	6
Grey Linnet	2	18	4	16	16	10
Redbreasted Linnet	2	16	4	14	16	8

	Mellow-ness.	Spright-liness.	Plaintive-ness.	Com-pass.	Execu-tion.	Dura-tion.
Twete or Mountain Linnet.	2	12	2	10	8	6
Greater Redpole	2	17	6	12	12	8
Lesser Redpole	1	14	4	8	8	6
Chaffinch	0	8	0	3	4	5
Bullfinch	4	6	6	6	6	5
Greenfinch	4	6	8	6	6	6
Yellow Bunting	0	4	0	2	2	2
Black-headed Bunting or Reed Sparrow.	0	4	0	6	6	2
To which we shall add, the Canary, though it is not an indigenous British Bird.	4	19	6	19	19	16

We shall endeavour to explain the meaning we
attach to each of these terms:—by mellowness—
we mean rich, full, deep-toned notes; by sprightliness—quick, gay, cheerful notes, generally expressive of joy; by plaintiveness—tender, pathe-
tic notes; by compass—forte and piano, also alto
and basso; by execution—the variety of notes in
a full, complete cadence; and by duration—not
the length of time a bird will continue to sing, but
the full length of its warble or song.

With regard to the teaching of birds, or their
acquirements in a state of domestication, the best
teachers never mix their lessons. If we wish to

teach them to pronounce a word, the same word
must be repeated till the bird has learned to articulate it; if we wish to learn them an air or a
sentence, it must be taught them by degrees: For
instance—if an air, whistle or play a small part
of it only, at a time; when that is learned, give
them more, and so on, till they have acquired, in
a perfect manner, the whole air; then a sentence,
or another air, may be taught them in the same
manner. In giving the lesson, always commence
at the beginning of the sentence or air, and repeat
it as far as the birds have already learned, adding,
as you find what has been taught them is fixed in
their memories; but never mix their lessons, never give them part of an air and part of a sentence
at one lesson;—let them learn one thing before
they are tried with another. To give more, at one
time, would confuse the birds, and what they acquired would only be a jumble. To teach birds
to sing in parts is very difficult; one is taught the
first or treble, and the other the second or tenor
part. The birds are taught in separate rooms,
and when each has learned its part they are then
brought together. At this period of their lessons
the birds require the greatest care; it is so difficult
to get them to sing together, and keep them to
their respective parts: Nor can the difficulty be

overcome, but by great skill, practice, and perseverance. All their lessons must be given just before they get their food.

We may here notice what appears to us a curious circumstance in the history of birds, at least as regards them in a state of domestication,—we mean their different dispositions; which vary as much in different birds as the dispositions of men vary from each other. Some birds are bullies, and quarrelsome, cruel to the females and young, indolent of song, careless, lazy, selfish, sullen, stubborn, and gluttonous: while others are docile and mild, kind, tender, and delicate to the females and young, lavish of song, industrious in building their little habitations, cheerful, and apparently happy. But what appears still more singular, is that the good qualities are generally found all in the same bird, and the bad ones all in some other. What can be more strikingly analogous than this to the characters of men? One man is industrious, honourable, and honest, kind and cheerful; another is idle, dissipated, and worthless, cruel, selfish, and stubborn. And as society despises the latter character, and esteems the former, so bird-fanciers, when they meet with a bird that is vindictive, stubborn, and sullen, they reject it, as they know it will have other bad qualities,—but prize a bird that is docile, cheerful, and kind, for they are sure it is to turn out an excellent bird. The dispositions, both of birds and men, may be ameliorated by education: But here the analogy ends—birds cannot improve their dispositions, but men may.

Much confusion has arisen in natural history, from the ambiguous terms employed by naturalists in describing the plumage of birds, the colours of which are often of importance in illustrating the species. Naturalists feel this, and also the want of a standard to refer to ; yet all agree that perspicuity is of the utmost importance in describing any object, and that the terms used in describing the colours of the plumage of birds, &c. ought to be clearly and distinctly understood.

Bewicke, in his Supplement to British Birds, takes notice of this; and regrets that no uniform standard has as yet been adopted to remove the ambiguous terms employed in describing colours; he particularly mentions the term cinereous, and says :—" Sometimes it appears to designate a distinct colour, with its various hues ; thus we have cinereous, pale cinereous, dark cinereous, &c. as it is synonimously, as from its etymology, it ought to be, if used at all, with ash grey; at other times it is confounded with many of the va-

rieties of brown and white."—"Language not less
vague is made use of, though perhaps not quite so
frequently, in regard to all the other principal colours; scarcely any two writers appearing to attach the same idea to the same diversity of shade."
To which may be added colour or tint.

He also mentions, that " the confusion to which
this has given rise, in ornithological descriptions,
has often made us wish, that naturalists would
adopt some uniform standard by which the subject of colour might be regulated, and, if possible,
fixed." The "nomenclature of colours" " of the
distinguished Werner, (as enlarged and exempli-
fied by Syme,) would seem to present the basis of
such a standard." We may also here mention,
that this nomenclature is highly approved of by
the celebrated naturalist Cuvier, in a letter from
him to Professor Jameson; and also, that it is
now adopted by the Professors of the University
of Edinburgh, and made use of by several eminent naturalists. Indeed, unless some universal
agreement, in regard to colours, be fully and distinctly understood, we may despair of ever seeing descriptive writing, of any sort, exhibiting
that accuracy, simplicity, and, at the same time,
correctness, which is so very desirable, and which
might thus be so easily attained.

For these reasons, in describing the colours of the plumage of the birds in this treatise, we have made use of Werner's Nomenclature of Colours by Syme.

This little treatise is accompanied with fifteen coloured engravings. The plates are admirably engraved, by Mr Scott of Edinburgh, from very correct and beautiful drawings done by an English artist; and the publisher has authorised us to state, that neither pains nor expense, on his part, has been spared to get the plates finely done and accurately coloured from nature; and, though we have no concern with this department of the work, yet we have no hesitation in stating our opinion, that it has been executed in such a manner, both as to elegance and accuracy, as cannot fail to give satisfaction to the public.

We cannot conclude, without acknowledging our obligations to several scientific gentlemen, to whom we are indebted for many useful hints, from which we have derived considerable advantage while employed in the arrangement of the present volume.

PATRICK SYME.

Edinburgh, 15th *July* 1823.

R. Scott Sculpt.

THE BLACKBIRD

BRITISH SONG-BIRDS.

When snow-drops die, and the green primrose leaves
Announce the coming flower, the merle's note,
Mellifluous, rich, deep-toned, fills all the vale,
And charms the ravished ear.   The hawthorn bush
New budded, is his perch ; there the grey dawn
He hails ; and there, with parting light, concludes
His melody.   There, when the buds begin
To break, he lays the fibrous roots ; and see,
His jetty breast embrowned ; the rounded clay
His jetty breast has soiled ; but now complete,
His partner and his helper in the work,
Happy, assumes possession of her home ;
While he, upon a neighbouring tree, his lay,
More richly full, melodiously renews.

Grahame's Birds of Scotland.

THE BLACKBIRD.

BLACK OUZEL, OR MERLE.

TURDUS MERULA; LINNÆUS.—LE MERLE ; BUFFON.

This bird, by the Scotch, particularly by their
poets, called the merle, is larger than the thrush.

As a song-bird, it is not equal to the thrush, but yet its notes are rich and mellow : it is esteemed an excellent cage-bird, but not for the aviary, as it pursues and harasses the other birds. In a wild state, it feeds on snails, earth-worms, spiders, and other insects, wild berries, &c. ; it is also very fond of cherries and pears, choosing always the best and ripest; it remains with us throughout the year, but in winter approaches houses and towns. We have seen it, during severe storms, come to the little gardens, (or back-greens, as they are called in Edinburgh,) quite close to the houses; nay, come to the very windows, and pick up crums of bread, &c. They are caught by limed twigs and snares of different kinds, but the birds taken from the nest are the best. To avoid repetition, we may here mention, that the young and old birds are reared and managed in the same manner as thrushes.—*See Thrush.* The nest is constructed nearly in the same places, nearly of the same materials, and, like the thrush, the female blackbird deposits four or five eggs, of a very pale verditer blue colour, faintly inclining to verdigris green, marked, particularly at the large end, with pale chesnut-brown spots. " The young birds are easily brought up tame, and may be taught to whistle a

variety of tunes, for which their clear, loud, and
spirited tones are well adapted." We have heard
a blackbird whistle "Over the water to Charlie,"
a jacobite air; "Dainty Davie," a Scotch air;
and an Irish air, without missing a single semi-
tone; but we think their own natural notes are
preferable to any taught air.

Description and Plumage.

This bird is about ten inches in length,—bill
straight,—upper mandible a little curved at the
point,—colour bright saffron yellow,—whole plumage rich velvet black, in some lights inclining to
bluish black,—inside of the mouth, edges of the
eyelids, and soles of the feet, pale saffron yellow,
—legs and feet dingy gallstone yellow. The female is blackish brown, paler on the breast,—bill
wood-brown, inclining to blackish brown,—legs
and feet the same. The young males, for the first
year, resemble the females, but after that their
bills turn yellow.

It is difficult to know a young male blackbird
or thrush from the female. The male, however,
is longer, and more slender. Choose the bird with
a large full sprightly eye, and slender towards the

tail. The last is a good criterion to judge by, perhaps the best that can be given, where both sexes are very like each other in plumage. We may remark, that, as the blackbird is more shy than the thrush, its haunts are more retired; for, though it occasionally builds in orchards, gardens, and hedges, we believe it prefers woods, copses, and distant shrubberies, and builds oftener than the thrush in evergreens, such as pines, cypresses, laurels, &c.

R. Scott Sculp.

THE THRUSH.

Published by John Anderson Junr. 55. North Bridge Street, Edinburgh. 1823.

> The thrush's song
> Is varied as his plumes; and as his plumes
> Blend beauteous, each with each, so run his notes
> Smoothly, with many a happy rise and fall.
> Sometimes below the never-fading leaves
> Of ivy close, that, overtwisting, binds
> Some riven rock, or nodding castle wall:
> Securely there the dam sits all day long;
> While from the adverse bank, on topmost shoot,
> Of odour-breathing birch, her mate's blythe chaunt
> Cheers her pent hours, and makes the wild woods ring.
>
> *Grahame's Birds of Scotland.*

THE THRUSH.

THROSTLE, OR MAVIS.

TURDUS MUSICUS; LINNÆUS.—LA GRIVE; BUFFON.

Of this genus there are four species that rank
very high as song-birds, *viz.* the thrush, or mavis,
sometimes called song-thrush,—the solitary or
moor-thrush,—the missel thrush,—and the blackbird, or black ouzel, already mentioned. All these

birds are remarkable for the richness and variety
of their song, but still more for the deep-toned
mellowness of their notes. In the northern countries of Europe, and even in France, the throstle
is said to be a migratory bird, but with us it remains, and braves our severest winters. Those
that migrate south in autumn from Norway, Sweden, &c. seem to shun the British isles; at least
they have never been observed to come in flocks
like their congeners the fieldfares and redwings.
Buffon says they are numerous during the vintage
in France, and, being very fond of grapes, they do
considerable damage to the vineyards, but, as soon
as grape-gathering is finished, they leave that
country for the south; and Sonini says he has
seen them in Egypt during the winter months.

These birds begin to sing early in spring. We
have often remarked their regularity in this respect, generally commencing in the vicinity of
Edinburgh about the fourteenth of February, unless in very severe storms of snow, and they continue to sing for nearly nine months of the year.

Perched on the topmost branch of a tree, with
their heads turned to the west,—there, in the stillness of evening, they may be heard pouring out, in
full, and deep-toned melody, their vesper hymn to

the setting sun; and again, at daybreak, saluting, with grateful warblings, that rising luminary, beginning so early as three o'clock; and we have also heard them chaunting delightfully between ten and eleven at night. In the middle of the day, when the sun is bright, and the air hot, they are generally mute, unless in cool situations among trees, whose thick foliage throws a glimmering light, " most like a lovely shade." In such a situation, and the ground level, in a fine calm summer day, their notes may be heard breaking the stillness, though the birds are more than a mile distant. These birds, when young, are very delicate, and require both care and attention to rear them; but, when reared from the nest, they make the best song-birds, and will sing nine months in the year. They are more tame, and even more healthy, than taught birds. The last, in a state of confinement, are often sullen, shy, and sing little, or do not let out their voice if overlooked; and thus deprives the master of much pleasing intercourse with his little proteges. Old birds and branchers are taken with nets, traps, &c.; but, as the thrush is not a rare bird, and the nest easily found, and, particularly, as the young turn out better songsters,—for these reasons we would give

the preference to a bird from the nest, and not
trouble ourselves with caught birds. The thrush
breeds early in spring, and has one, two, and
sometimes three broods in the year. The female
generally begins to make her nest in March, and
young thrushes have been seen about the middle
of April.

The Honourable Daines Barrington says, in his
scale of the comparative merits of singing birds,
(in which twenty is the supposed point of absolute
perfection,) that the song of the thrush has only
four of mellowness, sprightliness, compass, and
execution, while the hedge-sparrow's note possesses
six of mellowness, six of plaintiveness, four of
compass, and four of execution. Now, with deference to that gentleman's opinion, we think very
little of the hedge-sparrow as a song-bird, while
we rank the thrush very high in the scale of songsters. If we really understand the terms mellow,
sprightly, plaintive, &c. we should say, that the
natural note of the hedge-sparrow has *no* mellowness, four sprightliness, two plaintiveness, two
compass, and three execution; while the song of
the thrush has sixteen mellowness, six sprightliness, ten plaintiveness, twelve compass, and eight
execution. But this is probably a matter of opi-

-nion, and may arise from difference of taste and feeling; but we do certainly differ very much from that gentleman with respect to the merits of singing birds.—*See Introduction.* Thrushes may be taught to whistle tunes, and even to articulate words; but we are no great admirers of these forced conceits, for we prefer the natural wild notes of the little denizens of the woods and fields, to all these artificial acquirements.

The Nest.

The nest is composed of twigs, bent, moss, and grass; the inside neatly plastered with clay, and so compact, that, in rainy seasons, the eggs have sometimes been destroyed. The female lays four or five, rarely six eggs, of a very pale verditer blue, inclining to verdigris green, marked, particularly at the large end, with blackish-brown spots, some of the spots inclining to ink black.

To Find the Nest.

This bird displays little ingenuity in concealing its nest; it is, therefore, easily found, and thus becomes an easy prey to boys, cats, weasels, &c.;

but still it is not a rare bird, which may be explained by its having two or three broods in the year. Both male and female are employed in constructing the nest, which is placed in a hedge or bush pretty near the ground. We have found them in hedges, thorn bushes, and amongst the under branches of the silver and spruce firs. These last conceal it, for the branches must be lifted up or put aside before the nest can be discovered; but in hedges, &c. it is easily seen, as instinct compels these birds to build so early in spring that the foliage has not then had time to spread a green curtain round their mossy couch.

Treatment of the Young.

The young may be taken from the nest after they are ten or twelve days old. When they are brought home, put them in a box or round basket cage, upon fine dry hay, or fern; and keep them very clean; but if the weather is very cold, or the birds appear sickly, then the box or cage must be lined with flannel.

Feed them every two hours with minced flesh-meat, mixed with crums of loaf-bread; to which add one-third of bruised rape or hemp-seed. With care and attention they will thrive upon this; and, before they take the perch, put them in a large wicker cage, one half of the bottom of which ought to be covered with dry fern or clean hay, and the other half with fine gravel. They must have every morning a flat dish with water, large enough for them to wash themselves in. After they have done so, take it away, dry the cage, and then put in the dish with their food, already mentioned, and another dish with water for their drink. Old birds are managed in the same manner, only now and then they may get a spider or a few hog-lice, in Scotland called "sklaters," or a snail, or a slug. Thrushes are also very fond of the common garden snail, but they must have a stone in the cage to break the shell upon, which they will do themselves, and dexterously pick out the animal. We may mention here, that all birds, young and old, must be kept clean and dry, and their cages carefully cleaned every morning. The health of the birds depends on it.

Diseases.

Thrushes are generally very healthy birds; but, when affected with diseases, they are managed and treated much in the same manner as nightingales

and other soft-billed birds.—*See the article Night-*
ingale, Redbreast, &c. When the thrush is affected with the cramp, lay the bird gently on a
piece of flannel, and feed it with the mixture already mentioned, and give it occasionally a spider,
slug, or a few hog-lice. The food may be rather
more moist than when the birds are in perfect
health; and the water given them to drink ought
to have a little saffron or cochineal in it. When
they void too loose, their food, that is, the paste
made of flesh-meat, bread, &c. must be given
much thicker, and a little chalk must be crumbled
on the bottom of the cage. When their feet and
legs get clogged with any dirt or wet, take the
birds gently in the hand, and moisten the feet and
legs in tepid water, and wash all the dirt completely off.

Their food, in a wild state, consists of grapes,
currants, olives, cherries, pears, slugs, snails,
worms, spiders, hog-lice, and several other insects and wild berries.

Description and Plumage.

Length about nine inches, bill straight, upper
mandible slightly bent towards the point, and

notched near the point, on the under side of it.
Some strong hairs spring from the base of the bill,
particularly from that of the male, which is a
good mark, together with that of the brighter colours and sprightly look, to distinguish a cock-
bird from the hen. The eye is large, full, and of
an umber brown, or hazel colour; head, neck,
back, wings, and tail, pale umber brown, inclining
to hair-brown, in some lights having a tinge of olive-
green; breast pale wood-brown, slightly tinged
with rich ochre yellow, spotted with umber brown;
the spots shaped somewhat like an arrow-head,
lower parts yellowish white, also spotted with pale
umber brown, inclining to greyish black; legs and
feet pale chesnut-brown, inclining to wood-brown.

Song.

The song of this bird is really delightful. In a
calm evening, or fine still morning, he is heard to
most advantage, making the woods ring with his
melody, while echo, ready to catch the sound, repeats it, softened indeed, but still more sweet and
mellow; thus enhancing the pleasure of our morning and evening walks, by raising the heart with
grateful feelings to Him " who hath formed the

round world" and enriched it with so many won-
ders.   The low notes of this bird are particularly
rich and mellow, and the whole compass of his
song finely varied.   The note of the song-thrush
is peculiarly adapted to woods, groves, and gardens; but his song is generally considered too loud
for a room, unless placed in a distant apartment,
or outside of a window;—when the sound, thus
mellowed by distance, steals softly on the ear, it
is truly delightful.

As patriots guard their country from the steps·
Of some proud tyrant, and his lawless band,
Who, on the broad arena of the world,
Like gladiators, fight for prize and plunder,
" And spread destruction o'er a smiling land;"
So·dauntless guards the storm-cock his lov'd home,
His mate, his young, his nest, from prowling hawk.

Anonymous.

THE MISSEL THRUSH.

SCREECH THRUSH, OR STORM-COCK.

TURDUS VISCIVORUS; LINNÆUS.—LA DRAINE;

BUFFON.

This bird is the largest of the genus, and also
of European song-birds:—length about eleven
inches; breadth from wing to wing above eighteen inches. Its name is derived from feeding on
the berries of the misletoe; and, indeed, it is said
these berries are sometimes propagated after having

passed through this bird, and that the plant grows quicker and stronger after that process, than when it is engrafted by man. We certainly have seen it several times engrafted near Edinburgh, but always without success. The misle-bird is, except in size, so like the song-thrush in the colour of the plumage and general appearance, that we think a separate description unnecessary. The female begins early in March to construct her nest, which is commonly placed on trees thickly covered with grey lichen. It is composed of small pieces of sticks, rotten fibrous roots of shrubs, with the earth adhering to them, wool and coarse grass, interwoven with great art, and always covered on the outside with grey lichen or greenish-white moss;—the inside is lined with grass or hair. In this commodious, warm mansion, the female misle-thrush deposits from four to five eggs, of a pale but dingy flesh-red, marked, especially at the large end, with hyacinth-red, and deep orange-coloured brown spots. This bird, when intended for the cage, ought to be reared from the nest, and the young may be taken about twelve days old. Their treatment, and that of the old birds, as to food, diseases, &c. is the same as that recommended for the song-thrush. The missel-

thrush begins to sing early in January, if the weather is fine. In stormy weather, however, it is said
to have a loud screaming note, from whence it got
the name of storm-cock. It is a migratory bird,
and not so numerous in Britain as the common
thrush.

They are very quick in noticing birds of prey,
and very courageous and fierce. If a hawk approaches their haunts, which are generally orchards
and gardens, they instantly give the alarm by loud
screams, which appear to be understood by the
smaller feathered tribes within hearing of this war-
cry; for they soon assemble, and, headed by the
misle-thrush, boldly attack and drive off the in-
truder. He (the storm-cock) will singly attack
the jay, magpie, merlin, sparrow-hawk, and kes-
trel. It is worthy of remark, that every bird
seems to have a cry or note peculiar to its kind,
which it utters on the approach of a bird of prey,
and which note seems to be understood by birds
of different genera, from that which first gave
notice of a foe. On hearing it, each sounds the
alarm-note peculiar to its tribe, and, from all parts
within hearing of it, they congregate, and mu-
tually assist each other to drive away the common
enemy.

Song.

Amateurs and naturalists are much divided with regard to the merits of this bird's song. Some think it inferior to the song-thrush, others superior. That eminent ornithologist, Colonel Montagu, is of the latter opinion. He says, " The song of this bird is much louder, and superior to that of the throstle's. Perched on the uppermost branch of a tall tree, it sings while the female is making her nest, and during incubation, but becomes silent as soon as the young are hatched, and is no more heard till the beginning of the new year, unless the young are taken, or the female is destroyed: then it continues its song the whole summer." This experiment, he says, he has tried upon this and several other song-birds, and always found it invariable. With respect to the song of the missel-thrush, we ourselves know it to be loud indeed, but rich, and full of deep-toned mellowness.

This bird ought never to be kept in an aviary, or it will harass, perhaps kill, the other birds, and may destroy their eggs and young. In this respect, as well as in some of its other habits, we think it nearly allied to the butcher-birds, and that it forms the link between them and the thrushes.

THE SOLITARY THRUSH,

MOOR THRUSH, OR BROWN STARLING.

TURDUS SOLITARIUS; LINNÆUS.—LE MERLE SOLI-
TAIRE; BUFFON.

~~~~~~~~~~~~~~~

This is a rare bird in Britain, but a delightful
songster—very superior in richness, mellowness,
and compass, to any of the thrush genus. The
moor-thrush, or moor-starling,—for it is allied to
both, and seems to be the connecting link between
the stare and the thrush,—is a solitary bird, and
never seen in flocks—seldom even in pairs, except
during the breeding season. Its haunts are retired,
and far from the habitation of man—amongst
moors, heaths, and rocky ground, where it breeds,
and constructs its nest, which is placed in the
cliffs of rocks, or in birch bushes, or other natural
wood, on a branch near the root, and often close
to the rock. Sometimes, however, they build

among furze, (in Scotland called whins;) or, if an old edifice is near their solitary haunts, they, like the starling, will breed there in some hole amongst the ruins. From its nest being difficult to find, and from its being so rare a bird, it is seldom seen in a cage; but, by the true amateur of song-birds, it is highly prized; for, though its song resembles that of the throstle's, its notes, in all their qualities, much excel those of the song-thrush. Colonel Montagu says, " It is described as being common in France, Italy, and in the islands of the Mediterranean and Archipelago; that it frequents mountainous and rocky places, and, like the stare, it prepares its nest in old ruined edifices, church-towers, and other similar places, and lays five or six eggs." But two nests are never found near the same place.

The young are easily brought up, and repay the trouble by their sweet native song. They may also be taught to whistle, and articulate words, when confined. This species sings as well by candle-light as by day. Its food is principally insects, grapes, and other fruits. It is observed to change its abode with the seasons, coming into those parts where it usually breeds, in April, and departing in August.

Description and Plumage.

Both Montagu and Bewick agree that this bird, in size, figure, and general appearance, (colour excepted,) is more like the stare than the thrush. Its bill is pale, dingy hair-brown, inclining to yellowish-brown; and, like the starling's, is guarded by a prominent ring. It has a flat head, scarcely higher than the bill, very similar to that of the stare. Head, neck, back, wings, and tail, (the last somewhat forked,) are of an umber-brown colour—the edges and tips of the feathers are pale chesnut-brown, giving these parts a mottled appearance. The chin is greenish-white, tinged with yellowish-grey, and slightly mottled with very pale umber-brown. The breast and under parts pale umber-brown, and streaked, with yellowish-white legs, and feet yellowish-brown, inclining to chesnut-brown. These birds, young and old, with respect to their food, diseases, &c. are managed in the same manner as young and old thrushes.

" I can't get out !"—Poor bird ! has man's hard heart
Not field enough to wreck its cruelty
On fellow-man ?—but he, in wantonness
Of power, must seize on thee.—" I can't get out !"
Poor captive !   No, thy prison-bars are hacked
With instinct (Nature's) efforts to escape ;
While sweep the swallows pass in airy rounds,
Brushing with sportive wing thy prison-grate :
The while thy little heart beats strong, and pants
For Nature's gracious boon—sweet liberty.

Anonymous.

THE STARLING,

OR STARE.

STURNUS VULGARIS ; LINNÆUS.—L'ETOURNEAU ;

BUFFON.

THE Starling is a beautiful bird, and common in
Britain, but no songster. Its natural notes are a
harsh scream, and a chatter or twitter : but its capa-
bilities are great, for it soon learns to imitate the
notes of other birds ; and it may be easily taught
to whistle almost any simple tune, to articulate

R. Scott Sculp.

THE STARLING.

Published by John Anderson Jun.r 55. North Bridge Street. Edinburgh. 1823.

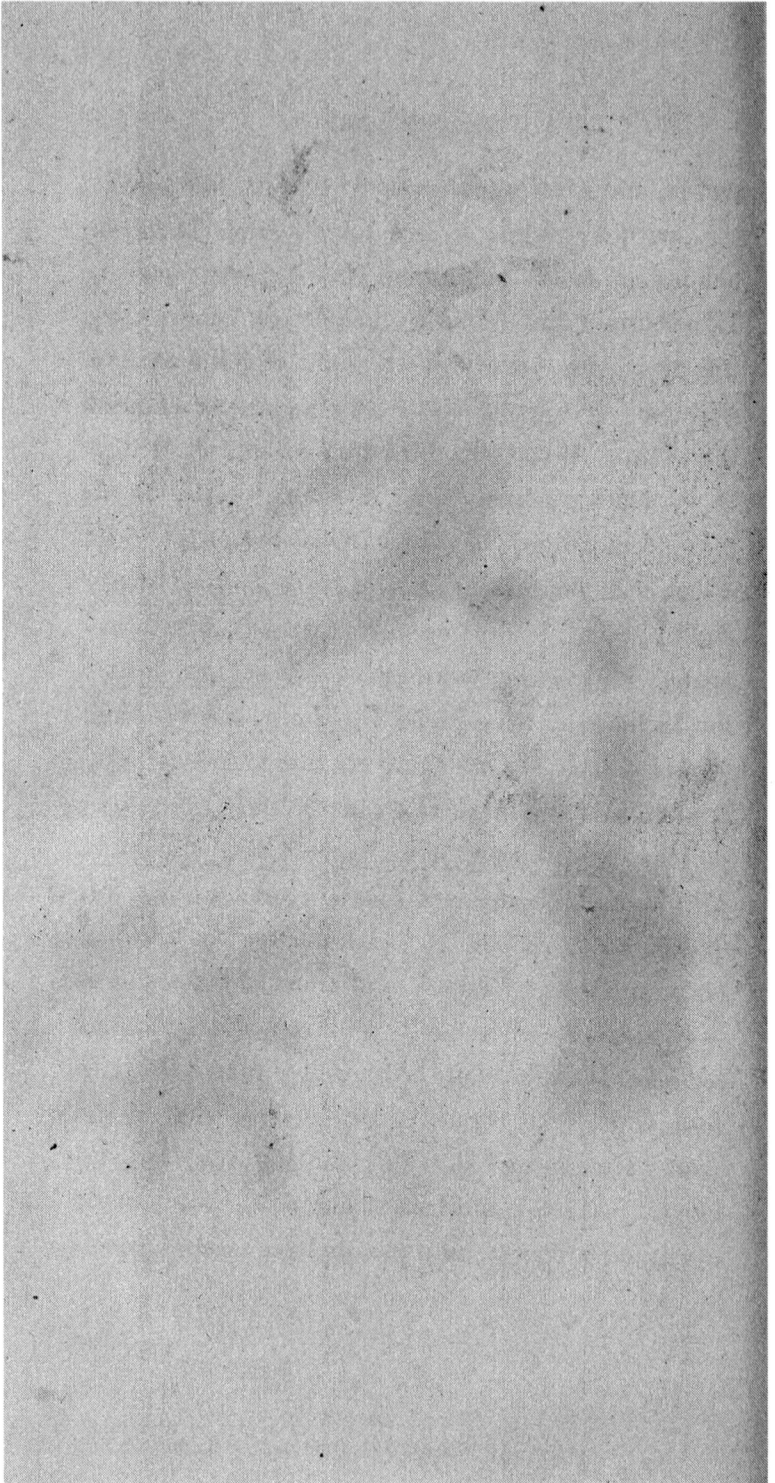

words, and even sentences. We went, one morning, with a friend, to see a collection of birds
belonging to a gentleman in Antigua Street,
Edinburgh; and among these were some very
fine starlings—one, in particular, which cost five
guineas. Breakfast was ready before we entered
the room. When the bird was produced, it flew
to its master's hand, and distinctly pronounced,
" Good morning, Sir—breakfast—breakfast." It
afterwards hopped to the table, examined every
cup; and, while thus employed, it occasionally repeated, " Breakfast—breakfast—bread and butter
for Jack—tea, tea—bread for Jack—pretty Jack
—pretty Jack." One thing we observed was this,
it often said the same word or sentence twice over
perhaps in imitation of the person who taught it.
M. Gerandin mentions, that a friend of his, M.
Thirel, which, when the bell rang for mass, called
on its mistress by name, and thus addressed her,
—" Mademoiselle, don't you hear the summons
for mass ?—Take your book, and return quickly to
feed your little rogue." Pliny says, that " the
young Cæsars had starlings and nightingales, docile in the Greek and Latin languages, which made
continual progress, and prattled phrases of consi-

derable length." It is also asserted that the kings
of Persia had starlings trained to hunt butterflies.

Starlings are found in great number from Norway to Siberia. Vast flocks of them assemble, on
the approach of winter, to migrate southward, and
have been traced to the Crimea, Natolia, and
Egypt; but in the temperate countries of Europe
they are stationary. According to Montagu, however, they migrate partially. He says, " We have
observed continued flights of those birds going
westward, into Devonshire and Cornwall, in hard
weather, and returning eastward as the frost
breaks up."—" But the vast flocks (says the same
author,) that are seen in severe winters, probably
migrate to this country in search of food, and return northward in the spring." Bewick mentions,
that, in the winter season, these birds fly in large
flocks, and may be known at a distance by their
whirling mode of flight, which Buffon compares to
a vortex. Another author remarks, that they have
a peculiar manner of flying, which appears to be
directed by a uniform and regular system of tac-
tics, and that each bird keeps constantly approaching the centre of the flock, while the rapidity of
its flight carries it beyond it. Thus, this multi-
tude of birds, united by one common tendency to-

wards the same point, flying to and fro, and cross-
ing each other in every direction, form a kind of
agitated mass, which appears to perform a general
revolution round itself, resulting from the particular movement of each of its parts. This method
of flight is attended with advantages. It protects
the flock from birds of prey, which, being embarass-
ed by numbers, incommoded by the flapping of
their wings, and stunned by their cries, is frequently obliged to abandon the tempting booty, without
being able to snatch the smallest part. This manner of flight protects them from the hawk; but
man profits by it, by means of limed lines.

These birds often assemble in flocks, in marshy
places, to roost; and the reeds are often much in-
jured by their weight. In Lincolnshire, quantities
of reeds are annually destroyed by them and rendered unfit for being used as thatch. The starling
is injurious to the labours of the husbandman;
yet his interest requires that they should be preserved, as they destroy vast numbers of pernicious
insects. Their food, in a wild state, consists of
snails, worms, grubs, beetles, and other insects;
but they will likewise eat wheat, rye, hemp-seed,
elder-berries, cherries, pears, grapes, and currants.

When tamed, their food is the same as that given
to blackbirds and thrushes; and they are, in every
respect, managed in the same manner as these
birds. Stares are fond of society, and will asso-
ciate, for want of better company, with crows,
rooks, jackdaws, pigeons, and redwings. They
take little trouble to provide a place for their progeny, frequently laying their eggs in the nest of
the woodpecker; and the latter sometimes pays
them back the friendly compliment. When starlings construct a nest for themselves, they merely
collect some grass and moss in a hole of a tree,
rock, or old tower, and upon this artless bed the
female deposits from four to six eggs, of a pale
bluish-green colour, inclining to very pale verditer
blue. It also builds in old castles, rocks by the
sea-side or inland, holes in high walls, and in
dovecotes. In the last they are sometimes taken
in great numbers. It is said they suck the eggs
of pigeons; but this we believe to be untrue.
They breed in May, and the female sits about 18
days. The young may be taken about 10 days
old, and should be reared in the same manner as
young thrushes. The starling is both docile and
quick, and will learn its lesson rapidly, and better

than if its tongue were cut—a cruel practice, and therefore very reprehensible.

Description and Plumage.

The length of this bird is nearly nine inches— crown of the head very flat, nearly equal in a line with the bill—nostrils covered by a prominent ring—bill sharp-pointed; in old male birds, of a pale saffron-yellow colour, brownish-black at the point—eye chesnut-brown—the whole ground-colour of the plumage bluish-black—the feathers on the head, neck, back, wings, and tail are edged with pale wood-brown—throat, and upper part of the breast, glossed with reddish-purple, and, on the breast and under parts, with deep duck-green, passing into Prussian blue—each feather on the head, neck, throat, and part of the back, marked at the point with very pale wood-brown spots, inclining to white; and, on the breast and under parts, the feathers are tipped with yellowish-white— legs and feet dingy orange-coloured brown. The young, till after the first moult, are of a sooty, umber-brown colour, somewhat like young blackbirds. In a wild state, they remain a long time dependent on their parents. The old female is very like

the old male in every respect, excepting in the
brilliancy of his plumage.   This is the best criterion to choose a male bird by that has moulted
once; for, before that period, the young cock may
be known by a black line under its tongue, which
they lose after the first moulting.

Now April starts, and calls around
The sleeping fragrance from the ground :
   Forgetful of their wintry trance,
The birds his presence greet.
But, chief, the Skylark warbles high
His trembling thrilling ecstasy ;
And, lessening from the dazzled sight,
Melts into air and liquid light.

        Gray.

SKYLARK.

LAVROCK.

ALAUDA ARVENSIS ; LINNÆUS.—L'ALOUETTE ;
BUFFON.

THE Skylark is generally admired, and esteemed an excellent song-bird, though its notes are
neither mellow nor plaintive. In this we differ
from the Honourable Daines Barrington, who
says, in his scale of the merits of singing birds,
that the skylark's note has four of both; but, in

sprightliness, we think he is right,—he says seventeen; and its song, in our opinion, surpasses all other warblers in cheerfulness. Whether this proceeds from the effect of its sprightly carol on the nerves, by touching those which raise the spirits, or from association of ideas, as bringing more immediately before us bright sunshine, rural scenes, green fields, gay landscapes, and nature in general, we cannot say. Perhaps the delight we experience, on hearing this cheerful warbler, may arise from both causes; but, to all who admire the wild warbling music of the groves and fields, and, more particularly, to the contemplative admirer of nature, the carol of the skylark does convey delight. The pleasure afforded to the mind by the melody of birds is pure and innocent. Does it not add greatly to the charms of a morning walk in the fields, to see the lark spring from his grassy couch, mount in air, and to hear him carol while he mounts, till he appears as a mere speck on the bosom of a white cloud, his thrilling notes still pouring over us, but from a height too vast for human eye to scan the airy minstrel.

The lark begins his hymn to the morning often before the sun is above the horizon. This is finely described by Burns:—

And when the lark, 'tween light and dark,
Blythe waukens by the daisy's side,
And mounts and sings on flitt'ring wings,
A wae-worn ghaist I hameward glide.

These birds commence singing very early in
spring. Their carol, at this season, is always a
prelude to pairing. We have heard them, in very
mild weather, towards the end of January, frequently in February; and in March and April
young larks have been found. Larks breed two,
three, and four times in the year; but birds of a
May brood are considered the strongest, and therefore supposed to turn out the best songsters. These
birds we should prefer for the cage next to those
reared from the nest. Pushers, by some, are considered to be as good as nestlings. " Pusher" is
a technical term for a young lark that has left the
nest; for the young always quit it, and run about
the spot several days before they are able to fly.
Pushers are caught by noting where the parents
light to feed them;—that observed, by running to
the spot, the young may be taken. Branchers are
full-grown birds, but that still have the nestling
feathers. They are ready in June and July, and
are caught with hawks and nets, trap-cages, &c.

and sometimes turn out very good birds. The old ones seldom tame. However, there are other kinds of bird-fanciers who think them excellent birds; we mean epicures. About Dunstable, larks are taken in nets, attracted to them by glasses called larking-glasses, which are fixed to a staff, and run out with a whirling motion. The larks, seeing themselves in the looking-glasses, fly down, when the nets are pulled over them, and many dozens, in this manner, have been caught at one pull. Larks, in a state of confinement, will sing eight or nine months in the year, and will live (with care) till they are 14 or 16 years old. The young ought to be kept by themselves, or only along with excellent song-birds, as they will catch the notes of any bird, whether good or bad. Well-taught singing larks bring high prices. In Sweden they are sold from three to six guineas each.

These birds haunt and build their nests in fields of grass, marshes, meadows, heaths; also in fields of young oats, barley, and other grain, according to the season. They make careless nests, of bent, coarse withered grass, &c. lined with horse hair, and, what is rather singular, we have observed the hair is generally white. The nest is commonly placed, (if early in the spring,) in a slight hollow, beside a clod or stone, to screen it from the cold, and always on the sunny side, *viz.* south or west. In this homely hovel,—for such it may be called, compared with the neat structures most of the small birds make for their young,—in this lowly dwelling the female deposits from four to six eggs, of a green colour, inclining to pale hair-brown, thickly freckled with spots of a deep hair-brown towards the large end, passing into umber-brown. She sits about sixteen days, and the young may be taken from nine to twelve days old.

Treatment of the Young.

The young ought to be taken about ten days old: if allowed to remain longer, they may be lost, as they are apt, (particularly in fine weather,) to run from the nest. When brought home, they are generally put into a clean straw or close wicker basket, with a lid, in which there is a hole. The basket should be constructed so as to allow the birds to get plenty of fresh air; therefore it ought not to be too small, and, for this reason, instead of a lid to the basket, we would recommend that a piece of fine muslin be tied over it, to prevent the birds from scrambling out. By not attending

to this, many fine birds have been lost. We ourselves had a nest of young swallows, who knew
our step so well when we went to feed them, that
they were always ready, at the door of a small
room where they were kept, to welcome our approach, and flew, the best way they could, to the
breast or hand to be fed; but, by not securing
them properly, they were at last all crushed by
the opening of the door in a careless manner. We
mention this to show how necessary it is to secure
all young birds properly, till they are caged or
able to feed themselves. The best food for young
larks is loaf bread, grated down, and boiled in
milk till it is pretty thick, to which add one-third
part of bruised rape-seed that has been boiled four
hours in pure water. After the water is strained
off, make the whole into a thick paste, and with
this feed the young birds every two hours. Each
bird may get from three to five pieces, about the
size of a small pea, on the end of a stick made for
the purpose; and, occasionally, they may get a
few small pieces of butcher-meat. After a week
they may be put into a larger cage, in the bottom
of which there ought to be dry hay, which must
be changed every day, and, in three weeks or a
month, the larks will be ready for the cage and

able to feed themselves. Their food, from this time, is loaf bread, minced egg, and hemp-seed. In the first week, or ten days, the hemp-seed may be slightly bruised, and a little of the soft food, *viz.* bread, milk, and whole rape-seed, may be put in a dish at one corner of the cage, that they may take it if inclined. Put a little fine gravel on the bottom of the cage; also a turf of grass or of fine-leaved clover. This must be given them every four, five, or, at most, six days; but the dry bread, eggs, and hemp-seed must be renewed every second day.

The male and female larks are so extremely like each other, that it is not easy to distinguish the cock-bird. The length of the claw of the hind toe, the white feathers of the tail and wings, and even the crest, are marks not to be depended upon. The best method, we believe, for judging of a song-bird, is to take the largest and longest bird, and put it in a cage, when, if a cock-bird, in less than a month it will "record:"—this is a technical word for warble or sing.

Description and Plumage.

The lark is elegantly shaped,—its length nearly

seven inches,—bill very pale, but dingy yellowish-brown,—head, neck, back, wings, and tail, umber-brown; but each feather is edged with pale wood-brown, inclining to pale yellowish-grey,—edges of the wings, and outer feathers of the tail, snow-white,—throat yellowish-white, marked in streaks with pale yellowish-brown, inclining to pale hair-brown,—breast, and lower parts, dingy greyish-white,—legs and feet pale yellowish-brown,—claws black,—hind claw very long: this last is a distinguishing mark of the lark genus. When alarmed, they raise the feathers on the top of the head, so as to form a kind of crest.

Song.

Of all song-birds, the lark is perhaps the most cheerful and sprightly. His natural song, though possessed of but little variety, seems endless, from the manner in which he renews it. He will often continue an hour upon wing, mounting till he is almost lost among the clouds, and then hovering, generally over the nest, chaunting all the time: He at last begins gradually to descend, warbling, as it were, with renewed energy as he approaches his mate. When about twelve or fifteen yards from the nest, his song ceases, he shuts his pinions, and drops like a stone to the earth. In a wild state, larks sing but little on the ground, but always carol when mounting in the air or descending. Should a bird of prey, however, appear, they become instantly mute, and, in this case, they will drop to the earth from a great height. When near the ground they flutter their wings to break the fall, and at the same time take a skimming or slanting flight for a few yards, according to circumstances. This slanting flight is a general practice with them when danger is near their nests. They never descend immediately to the nest, but a good way from it; they then fly close to the surface of the ground for some yards, and afterwards creep cunningly to it through the grass.

By winding Ayr, or Lugar's stream,
Where thrush and merle the green woods throng,
Oft have I paused at day's last beam,
To hear the wood-lark's plaintive song ;
Now swelling on the evening breeze,
Now lost upon the list'ning ear,—
Soft mingling with the rustling trees,
Or with the streamlet murm'ring near.
It seem'd as if some dirge it sung
Of other times and happier hours ;
So sad, so sweet, the cadence rung
Amid those wild and lonely bowers !

Anonymous.

THE WOODLARK.

ALAUDA ARBOREA ; LINNÆUS.—L'ALOUETTE DE

BOIS ; BUFFON.

THIS charming bird, so deservedly esteemed for
its song, is, by some, thought nearly equal to the
nightingale, while others prefer it to that bird.
Its warble certainly possesses most of the qualities

R. Scott Sculp.ᵗ

WOODLARK.

Published by John Anderson Junᵣ 55. North Bridge Street. Edinburgh.1823.

of the nightingale's notes, but it is neither so mellow nor so plaintive, though equally rich and expressive. Like the skylark, they sing while flying, and warble over the nest for a long time,
without any apparent break in their song. This,
we believe, is the only species of lark that sings
when perched on a tree, from which it pours its
notes in rich and flowing melody. We think its
warble from a tree is more full and tender than
when it sings in the air; but this may arise from
the bird being then nearer us, or from the stillness
of the morning and evening, at which time he generally pours forth his softest notes to his mate.
Perched on a limber branch or bending spray, almost over the nest, he salutes his assiduous mate
at day-break; and again at evening, to a late hour,
he serenades her in even more soothing strains.
From its singing so late in the evening, it has
been called the Scotch nightingale. The woodlark frequents copse wood, and banks that slope to
the sun and that are covered with briers and brambles. The nest is constructed, either at the root
of a tree, or among brambles and wild rose-bushes,
generally in a tuft of long grass; it is formed of
bent and dry grass, lined with thistle-down, wool,
and hair, rather shallow, and carelessly made. In

this the hen drops four or five eggs, of a dingy reddish-white colour, sometimes brownish, tinged with very pale yellowish-brown, speckled with pale and deep orange-coloured brown spots; but the eggs vary in colour. The hen sits about fifteen days, and the young may be taken about ten days old; but, as this is a delicate bird when young, and the nest not easily found, caught birds are rather preferred for the cage.

The young are reared much in the same manner as skylarks, but no milk is mixed with their food. They should be fed with loaf-bread, egg, and flesh-meat, about equal parts of each, all finely minced and mixed together, and moistened with a little water. In every other particular the young are lodged and treated in the same manner as skylarks. The young, when ready for the cage, and old birds, are fed with the same paste already mentioned, but, occasionally, there ought to be mixed with it one-fourth part of hemp or rape-seed. In hot weather they must have a flat dish to wash themselves in, always after which the cage must be dried, and the bottom of it strewed with fine gravel. If unwell, give the birds ants' mould, with the insects and their eggs in it. Should they void loose, give them a little grated chalk. In every respect, as we before stated, they must be managed in the same manner as skylarks.

The woodlark sings about nine months in the year; nay, it is sometimes even heard in January if the sun shines. They will learn the notes of the nightingale, and blend them with their own; and a bird thus bred will sell at a high price.

They are caught, like the skylarks, with hawk and nets, and sometimes with trap-cages. When taken, tie their wings, and put the birds in a dark place, and forbear feeding them for some time. They are fed with nightingale's food, a few ants and some meal-worms being first put amongst the gravel on the bottom of the cage, and on or near their food, to tempt them to eat. When tamed, they require nothing but care, to be kept clean and dry, and fresh food and water given them every day. To each bird give about a dessert spoon-ful of the bread, eggs, and minced flesh-meat, mixed with bruised hemp, rape, and maw-seed, or the seed may be given on a separate dish, let-ting them choose for themselves. The best birds are taken in September, and before they pair in spring;—branchers may be caught in June or July.

The male and female are very like each other;

but the head of the cock-bird is very flat and full
behind : this, and the brightness of the whitish
streak proceeding from the bill to behind the eye,
forming a curved line over it, and almost meeting
behind the neck, are the best marks to distinguish
the song-bird from the female.

Description and Plumage.

This is a very elegant bird, and the plumage,
though plain, is harmonious and prettily varied.
Length near six inches,—bill pale yellowish-brown,
darkest at the point,—eye chesnut-brown,—upper
part of the head, neck, back, wings, and tail, pale
umber-brown, streaked down the middle of each
feather with a very dark mark of the same colour,
—spurious wing edged with dingy yellowish-white,
—a streak of the same passes over each eye, extending pretty far back,—throat, breast, and under
parts very pale wood-brown, passing into yellowish-white towards the tail,—the throat and breast
spotted sparingly with umber-brown,—the two
outer feathers of the tail snow-white,—legs and
feet dingy yellowish-white,—hind claw long; but
neither the hind claw nor tail so long in proportion
as that of the skylark's.

Song.

The natural notes of the woodlark are really charming, and full of melody: they are rich, expressive, and sweet, particularly when the hen is occupied with incubation; for at that time the cock-bird cheers her incessantly with his song, but, like all other song-birds, he is mute the moment the young are hatched.

The natural warble of the woodlark seems composed of part of the notes of the nightingale's, blackcap's, and skylark's warbles. Though not equal to the two first in mellowness, nor to the last in cheerfulness, yet it is a delightful, expressive, sweet, and varied song. It is a bird of great power and perseverance, and therefore excellent for being put beside young canaries, goldfinches, chaffinches, titlarks, &c.

THE FIELDLARK.

ALAUDA CAMPESTRIS; LINNÆUS.—LA SPIPOLETTE;

BUFFON.

THIS is another good song-bird of the lark kind.
It is about the size of the woodlark, and so like
it in plumage that a separate description is unnecessary; only the tail is longer, and the hind
claw more curved in this bird than in those of the
woodlark, and its song, though sweet, is not equal
to that bird's.

Montagu says, " This bird has often been confounded with the titlark; but the base of the bill
is broader, and the hind claw shorter and more
curved, marks that cannot be mistaken. Its song
is vastly superior to the titlark's, though something similar. This it delivers from the branch of
a tree, or on wing, as it is descending to the ground."
It makes its nest amongst the high grass or green

wheat. The fieldlark is fed, reared, and managed, in the same manner as the woodlark, and, being rather uncommon, and a sweet song-bird, is much admired by amateurs. We may here mention, that Bewick says, " We have occasionally met with another bird of the lark kind. It frequents woods, and sits on the higher branches of trees, from whence it rises singing to a considerable height, descending slowly, with its wings expanded, and tail spread out like a fan. Its note is full, clear, melodious, and peculiar to its kind."

THE TITLARK.

ALAUDA PRATENSIS; LINNÆUS.—L'ALOUETTE DE
PRES; BUFFON.

THE Titlark has been often confounded with
the fieldlark and pipet; but it is smaller than either of these birds; it is, besides, very common,
and frequents barren situations. In Scotland, it
is almost the only bird found upon the vast extended heaths amongst which it breeds.

We have often seen them upon Arthur's Seat
and Salisbury Crags, near Edinburgh; likewise
on the Calton-Hill, on which is now built part of
that city; and there, before the public walks, &c.
were made, we have found its nest. This bird,
like most of the lark genus, sings on the wing,
springing up, hovering a little, and then descending slowly, warbling till it reaches the ground,
which it does with a kind of sweep and a jerk of
the tail as it alights. Its natural song is sweet,

but short; but, by proper care, it may be made an excellent song-bird; and, after being taught, it may be put, with great advantage, beside young canaries, goldfinches, chaffinches, &c. who will readily learn its notes. The nest of this bird is made of coarse grass; sometimes a little moss is added, and lined with fine grass and horse-hair; it is placed in tufts of grass, at the roots of furze, or close to a bush or stone near the ground. The eggs, four to six in number, " vary considerably in colour, some being of a dark reddish-brown, others whitish," thickly speckled with reddish-brown, or pale orange-coloured brown spots. The hen builds her nest in April, and the young may be taken in May. The young are reared, and the young and old fed and treated in the same manner as the sky-lark, only now and then they may get a few ant-eggs, and two or three meal-worms, particularly when moulting; and also a little saffron may be put into the water when they appear husky: all birds are liable to this, which proceeds from cold. Often, though the sun is bright, the external air is chill; and, if the window is open, a stream of cold air rushes in, which is apt to produce this malady: it is analogous to sore throat in the human species. To cure it, they must be kept warm and dry, and saffron or a little liquorice-root put into their drink.

Description and Plumage.

This bird is in length about five inches and a half,—of a slender make,—bill wood-brown, inclining to black at the tip,—eyes umber-brown,—over each eye is a streak of pale yellowish-grey, inclining to white,—the upper part of the plumage is not unlike the skylark, but darker on the back, where it passes into hair-brown, slightly tinged with olive-green;—the breast is very pale, but dingy straw-yellow, beautifully marked with brownish-black spots,—the under parts pale, but dingy ash-grey, inclining to white,—obscurely streaked on the sides with very pale clove-brown,—tail umber-brown,—outer feathers white,—legs pale dingy ochre yellow,—feet and claws pale yellowish-brown. The male and female are very like each other, and the only way of distinguishing the cock-bird is to hear it record, or warble its song.

And shall ne'er Scotia hail thee, Philomel,
Enchantress sweet! ne'er hail thy vesper strain;
Or hear thy melting notes steal through her vales
In mellow murmurs?   Say, then, warbler, say,
Why her sequester'd glades, her flow'ry glens,
And briery brakes, for thee no charms display!
They smile as sweet as those in southern climes,
Through which thy little pipe pours melody.
Then, warbler, come! it is a land of song,
And join thy minstrelsy.   Poor Scotia's sons,
Though rude their northern harp, shall welcome thee
With strains as soft as breath of whisp'ring lute,
Or elfin magic lay on zephyr borne.
What though no myrtle loads the air with fragrance,
The lowly violet scents the gale as sweet;
Then, Philomela, come! to Scotia come!
And, when the yellow moonbeam steals along,
Pour through her woods thine own soft plaintive notes.

Anonymous.

THE NIGHTINGALE.

(POETICALLY) PHILOMEL.

SYLVIA LUSCINIA; LATHAM.——MOTACILLA LUSCINIA;
LINNÆUS.——ROSSIGNOL; BUFFON.

THE Nightingale, as a song-bird, is deservedly
esteemed one of the sweetest, most powerful, and

certainly the most melodious of E
In Asia and Africa he is equally p
unrivalled excellence he may, p
the first songster in the world.
may have their favourite song·
prefer the note of the woodlark,
the canary.; but it is well known
is decidedly in favour of the nig
different countries of Europe wh
found, they are only partial visi
ring a very few of the spring and
Nightingales arrive in Englan
March or the beginning of April
wards the end of August, and h
broods in the season. These]
prized by the Romans. Pliny s
bird, which, for fifteen days and
thickest shades, continues her n
mission, deserves our attention a
surprising that so great a voice
small a body !—such perseveral

unexpected transition,—now seeming to r
same strain,—then deceiving expectatio
sometimes seems to murmur within herse
deep, swift, trembling,—now at the top,
dle, and the bottom of the scale. In shor
little bill seems to reside all the meloc
man has vainly laboured to bring from
of musical instruments. Some even see
possessed of a different song from the
they contend with great vigour; and the b
come is seen only to discontinue its song
life."

In all countries where this delightful
is found, it is held in the greatest estima
is celebrated in the lyrical and amatory
the Persians. Hafiz, (the Burns or N
Persia,) in his odes, has sung its fame i
sweet, tender strain, that it can only be eq
the melting melody of the bird whose p
sings. In that country, the poetical att
of the bulbul (or nightingale) to the ro
teemed an emblem of the purest love

victory to man himself. " An intelligent Persian (says the late Sir William Jones, in his Dissertation on the Musical Modes of the Hindus,) declared he had more than once been present when a celebrated lutanist, surnamed Bulbul, (the nightingale,) was playing to a large company, in a grove near Shiraz, when he distinctly saw and heard the nightingales trying to vie with the musician,—sometimes warbling on the trees,—sometimes fluttering from branch to branch, as if they wished to approach the instrument,—and at length dropping on the ground in a kind of ecstasy, from which they were soon raised, he assured me, by a change in the mode."

M. Gerardin happening to saunter in the Jardin des Plantes, at Paris, in a fine spring evening, his ear was regaled with the melodious accents of two nightingales. He instantly returned the compliment by some passages of tender airs on his German flute, when the feathered musicians approached him,—first in silence,—but, after listening for a while, they sang in unison to his instrument, and soon surpassed its powers. On raising his key, first one-third, subsequently a whole octave, they shrunk not from the challenge, and acquitted themselves in such a style, as, by M. Gerardin's own confession, to merit the wreath of victory.

Two gentlemen, walking in the environs of London, close by a copse, were struck by the sweetness and wonderful richness of the notes of a nightingale, which sung in a strain superior to any they had ever heard, though in the habit of hearing many. The place was a valley, with brushwood on one side; all the notes sounded double, the effect of an echo, which may account for the great richness of the bird's song. A woodlark, perhaps attracted by it, flew, perched near the spot, and endeavoured to outvie the nightingale; another woodlark immediately appeared, and placed himself scarce an hundred paces from the first, and under these happened to be another nightingale. These four birds warbled out their notes, for nearly an hour, in strains the most delightful,—yet with such seeming contention, that it was amusing to see the keenness of these pigmy rivals. The larks gave way first, and went off; the nightingales soon followed. A short time after, one of the gentlemen, while humming a tune, was astonished to hear himself accompanied by one of the nightingales, which had returned unnoticed; and, in a few minutes after, the other nightingale reappeared, and did the same. The accuracy, management, and flexibility of voice displayed by these

little creatures were wonderful in keeping in unison with the gentleman's song, but which they
did in the most perfect manner.

Woodlarks appear to be their principal compe-
titors for musical fame, though they will contend
in the same manner with any other songster—and
with such vigour, spirit, and perseverance, that
they seem rather resolved to die than be outdone.
In a domestic state this bird has been known to
contend with a woodlark for half an hour, till,
from exhaustion, he has been like to drop from his
perch—and probably might have done so, but
that the woodlark was silenced by the other's un-
conquerable spirit. The formation of this little
warbler's throat enables it to continue its song
longer than any other bird, without taking breath:
When that becomes necessary, it does it with all
the skill of the most experienced opera singer,
the pause not being perceptible;—again the strain
flows on, mellow, sprightly, rich, and plaintive.
One cannot wonder, therefore, that the nightingale
is esteemed the first of song-brids, and in all countries sought after with the greatest avidity.

In most of the principal cities of Europe there
are people who make a profession of rearing them,
the birds bringing very high prices; but this pro-

cess, and that of their education, either for the aviary or cage, requires skill—and, being attend- ed with much trouble, can only be accomplished by kind treatment and unremitting attention : yet many die before they are brought to a state of domestication, not so much through sullenness as from the delicacy of their constitutions.

The bird-teachers of Rome, Naples, Florence, Vienna, Paris, and London, are famed for rearing these birds; and the intelligent traveller, Dr Clarke, mentions Moscow, where, he says, night- ingales are heard during the night, making the city resound with the melody of the forest. Its song in Persia, Natolia, and Greece is said to be much finer than in Italy.—The Italian birds are more esteemed for this quality than those of France, and the French are considered better than the English birds; even in England, one county is said to produce a finer song-bird than another, though the districts join. Thus ama- teurs, in London technically called " bird-fan- ciers," prefer the nightingales of Surrey to those of Middlesex. This bird is therefore supposed to be heard in the greatest perfection in the east, and said to decline in the sweetness and richness of its vocal powers as it migrates north

and west. This may be true; but we shall state
what came under our own personal observation
with regard to the comparative excellence of the
song of these birds on the continent and in Scotland.

In 1802, being at Geneva, at the residence of
a friend about three miles from the town, in a
quiet sequestered spot, surrounded by gardens
and forests, and within hearing of the murmur of
the Rhone :—there, on a beautiful still evening,
the air soft and balmy, the windows of the house
open, and the twilight chequered by trees, there
we heard two nightingales sing indeed most delightfully,—but not more so than one we heard
down a stair, in a dark cellar, in the High Street
of Edinburgh !—such a place as that described in
" *The Antiquary* ;" no window, and no light admitted but what came from the open door, and
the atmosphere charged with the fumes of to-
bacco and spirits : it was a place where carriers
lodged, or put up,—and the heads of the porters
and chairmen, carrying luggage, nearly came in
contact with the cage, which was hung at the foot
of the stair-case :—yet even here did this bird sing
as mellow, as sweet, and as sprightly as did those
at Geneva. We have often stopped to hear it,

and listened with the greatest pleasure; and, as
the pieman passed with his jingling bell, a sound
now seldom heard in the streets of Edinburgh,
the bird seemed more sprightly, and warbled with
renewed spirit and energy ! So much for the supposed superiority of continental to British birds.

In all parts of Europe nightingales are migratory, retiring to Asia and Africa from the severi-
ty of our winters; but in India, China, and Ja-
pan, they are permanent inhabitants. In the
islands of the latter empire these birds are an article of commerce, and are sold at a high price.—
Sonnini says, he has traced their passage to the
islands of the Archipelago, and in winter has met
them on the fresh and smiling plains of the Delta.

This bird has a sort of jerk with its body and
tail, somewhat like the redbreast; and in manners
nearly resembles that bird, except in approaching
the habitations of men, being more like those redbreasts that retire to the forests to breed. It is a
shy, solitary bird, and, like the redbreast, will allow no other of the same species to approach its
haunts: their food is the same, their nest and
their manner of building nearly the same; their
little dwelling is concealed with equal ingenuity;
they affect the same situations for incubation, and

E

the nest may be discovered in the same manner as
that of the robin's, for, at no great distance from
the nest, the cock, perched on the branch of some
favourite tree, may be heard, chaunting to his
mate his vesper notes in sweetly-soothing melody,
thus cheering her during the toil of incubation.
The nightingale is not unlike the hen redbreast, but
not such a bunchy bird, being more slender, neat-
er, a little longer, and the colour of the plumage
rather lighter. They are not so widely diffused
in Britain as the redbreast and blackcap,—the
birds that resemble them most in the mellowness,
richness, and tenderness of song: But they affect
the same places, such as retired situations amongst
copses, thickset hedges, entangled brushwood, in
solitary dells, near chalk-hills, where a pebbly
stream steals through bushes and underwood.
Their haunts, at their first coming in spring, are
the roots of hedges, briers, brambles, &c.—There
they are screened from the cold, and there also
they find their favourite food. This is the best
time for taking old birds, *viz.* before they pair ;—
when taken after that period they are apt to die.
It is said the males in number greatly exceed the
females; but this opinion Colonel Montagu dissents
from, giving, as a reason, that, as the males arrive

about ten days earlier than the females, and consequently at this time none but males are caught, he supposes this circumstance will explain the apparent disparity.

Of the Nest and Eggs.

Nightingales breed in May. The second brood (if they have two,) is ready about the middle of July. The nest is artfully hid under the tangled roots and lower branches of old trees, or thickset hedges, where brambles and briers, matted and interwoven, present a thorny *chevaux de frize* to all intruders. It cannot be found without tracing the bird to the spot, and even then not without great difficulty. It is composed of dried leaves, withered grass, and straw and moss bound together with fibrous substances, and lined with soft hair and down. The eggs, four or five in number, are of a pale hair-brown, inclining to broccoli-brown, somewhat resembling a pale-coloured nutmeg. In this climate, it is said, the eggs do not all come to perfection.

To find the Nest.

It is found in the same manner as that of the redbreast, *viz.* by watching the cock. If he appear shy, endeavour to tire him out, but keep still: he may also be thrown off his guard by being lured with meal-worms: Stick some of these on shrubs or branches where the bird may see them; or, in some favourable spot, turn up the earth, placing thorny twigs in it, and on these put some of the worms: The new-turned earth will attract him, and, when he sees the tempting bait, he will seize it for his young; trace where he goes, listen, and the chirping of the nestlings may be heard, as they always make a great noise when the parents appear. If the nest be found, and the young not ready for taking, (for they should be nearly as well feathered as the parent bird,) do not touch them, in case they should scramble over the nest or lest the instinct of the old birds should lead them away.

Of Old Birds.

From the last week in March to the middle of April old birds may be taken, but not later, as then they begin to pair, and are apt to pine and die when taken from their mates. They may be caught with cage-traps and limed twigs. To lure them to the twigs, stick meal-worms to them; but the cage must be baited with a little mould from an ant-hill, upon which strew ant-eggs and a few meal-worms. Be sure to turn up the earth the breadth of a foot or so round the cage, and the fresh mould will attract the birds to the spot;— but some consider the nightingale's trap the best method of catching them: It is made of a circular, hollowed piece of wood, about a foot in diameter, with a circular wire the size of the trap, to which is attached a green silk net; there is also a watch-spring, a string to hold up the trap, and a little cork: it must be baited with meal-worms, which, when the bird seizes, the net falls down and secures him. The earth round this must likewise be turned up, and the trap placed as near their haunts as possible, particularly about the spots where they are seen to feed.

Branchers.

Young birds are called branchers from the time of their leaving the nest until their departure in autumn. The birds of the first brood are the best, and may be taken in July. Those of the second are often too weak to stand the winter: they are caught in the same manner as old birds, and both old and young, when secured, must have the tips of their wings tied with packthread, (but not too tight,) to prevent their dashing themselves against the sides of the cage. Put them into a nightingale's back cage, and keep them quiet; if in a common cage, darken it with a piece of cloth, and let them alone for about two hours till they become settled, after which offer them food; if they are sullen, and refuse to eat, they must be " crammed,"—a term used by bird-fanciers; *viz.* take them gently in the hand, open their bills with a thin stick made for the purpose, and put in a pellet (about the size of a small pea,) of nightingale's food, which consists of fine-grained butcher-meat, or sheep's heart carefully freed from fat, minced very small, and mixed with a little of the yolk of a hard-boiled egg, and this they ought to get every two hours until they are reconciled to captivity. To entice them to eat, put a little of their food in the pan, and on the bottom of the cage strew some mould from an ant-hill, amongst which put some ants, ant-eggs, and meal-worms: If these are picked out, put more, and continue this till the birds are so far tamed as to feed themselves and take both food and water cheerfully from their glass and pan; then their wings may be untied.   Some think the best way of rendering the branchers tame, is to confine them in a cage surrounded with boards on all sides except one, which is covered by a green silk net, through which the food is given.

Treatment of the Young.

Many prefer the young from the nest to either old birds or branchers, as they are easier tamed, and get more attached to those who feed them than caught birds do.. To rear the young is indeed a very delightful amusement, though it certainly is attended with trouble, and requires great attention, at least for ten or twelve days; but the attachment of these little warblers amply compensates for the toil bestowed upon them.   Young birds ought to be about twelve or fourteen days old before they are taken. `If brought away sooner, they no doubt take food more readily; but, from the delicacy of their constitutions, many die.   Let them therefore be well fledged, if it can be ascertained; the best time for taking them is a few days before they are able to fly.   Take the nest along with the young; put them into a clean, airy cage, in a quiet place, and cover it so as to produce an effect similar to a dull twilight: Noise and bustle frighten them; fright produces fainting, and this may occasion fits which often end in death. To render nightingales tame, be gentle, tender, and keep them very quiet, particularly at first. When first brought home, they sometimes refuse to take food. Care and perseverance are then necessary; but the same method as that recommended for old birds and branchers, will bring them round, *viz.* open their bills and put in pellets of the paste or mixture of sheep's heart and egg, with now and then an ant's egg or meal-worm. When they take the perch, strew ant's mould on the bottom of the cage, which last ought to be cleaned and the gravel renewed every day. When they are able to feed themselves, put some of their food (the paste) into a pan or china dish, and some pure water into another, and be sure to do this daily. The dish is preferable to the pan, as the latter is apt to rust. If they droop or appear dull, or lothe at their food, give then occasionably a spider, an ant's egg, or a meal-worm. All young birds, while unable to feed themselves, must be fed very early in the morning, and regularly every two hours during the day.

Nightingale's Food.

Their common food is a thick paste or mixture made of sheep's heart or fine-grained butcher-meat, (freed from all stringy substances, skin, and particularly fat,) mixed with the yolks of hard-boiled eggs, the whole minced very fine and moistened with a little water; sometimes sweet biscuit is added, but of this we do not approve. Nightingales also occasionally get figs chopped small, spiders, ants' eggs, loaf sugar, flies, soft smooth grubs (they will not touch a hairy one,) and the caterpillars of butterflies, moths, and hawk-moths; but these are only given when the birds are not in good health. Some even give cheese; but this we consider detrimental to all birds. Many feed their nightingales with German paste, an article will known to amateurs or bird-fanciers, and to bird-catchers. Others give their birds a paste made of particular herbs and the minced heart; but this has proved fatal to many a fine bird. We would therefore recommend, as the best food for these birds, the common " nightingale's food," already mentioned. Give of this paste to each bird, every day, about the size of a nutmeg, which must be fresh made and of sweet materials.

E 2

In winter, parboil the heart or butcher-meat before adding the eggs; only, instead of water, use a little of the liquid in which the heart is boiled. To keep the liquor pure, all the fat must be carefully picked from the heart before it is boiled.

Diseases.

In a state of confinement the diseases of small birds are very similar and the treatment much the same. Some, however, are more liable to maladies than others; and we rather think the soft-billed birds are not so hardy as the finches. For this reason, and because of the great delicacy of the nightingale's constitution, disease in that bird must be taken in time, lest it prove fatal. When kept long in a cage, they are liable to gout, that is, the legs and feet swell, and the claws drop off; for this, rub the parts affected with fresh butter or lard.

These birds always void loose; but if more so than usual, take some hemp-seed bruised or ground, in quantity as much as will lie on a sixpence, and mix it in their common food, to which may be added a little loaf sugar; or give them a small portion of prepared chalk (*creta ppt.* of the shops.)

If the birds get very thin, give them chopped figs ; if too fat, a little grated rusk, known by the name of tops and bottoms. When they appear dull, bunch out their feathers, or put their head under their wings: give them two spiders, a few ants or ant-eggs every day until they recover their spright- liness and their feathers again lie close and sleek; for this also a little saffron or a slice of liquorice- root may be added to the water.

As cleanliness is absolutely necessary to preserve these birds in health, put a flat dish with water on the bottom of the cage, that they may wash them- sevles every morning; after which the dish should be removed and the cage properly dried. It some- times happens, unavoidably, that stringy parts of the heart, or pieces of skin, are chopped into their food, and these may get entangled about their tongue. Should that take place, it may be detected by their gaping and an effort to vomit; in that case, take the bird gently, open its bill, and, with a needle or hook made for the purpose, pick out the obstructing matter, or death will ensue; after which, give the bird a few spiders. This acciden- tal evil may be easily distinguished from the dis- ease called " the gapes," which is, when the birds open their bills frequently as if for breath, and

occasionally sneeze as if to eject something from
their nostrils; then it is vermin in their throat,
and for this complaint put a little saffron or liquorice-root and a small piece of white sugar-candy
into the water given them to drink.   Should they
be troubled with the pip, which often attacks them
while moulting, take a needle and puncture the
part, and gently press it; after which rub it with
a little lard or fresh butter.   The pip is a swelling
of the gland, which secretes the oil with which
birds dress their feathers; it is situated just above
the tail.

Description and Plumage.

Length, six inches and a half;—make, slender.
In appearance, active and sprightly;—eye full,
large, clear; of an umber-brown colour or dark
hazel; bill, umber-brown at the base, tinged with
dingy primrose-yellow; head, neck, back, and
wings, light yellowish brown, tail a little darker;
the middle feathers of the tail and primary quills
the same tint, but slightly inclining to reddish-
brown; throat, breast, and under parts yellowish-
white; at the side of the breast, near the spurious

wings, tinged with pale wood-brown; legs and feet
dull flesh-red; claws brownish-black.

Of their Song, and Musical Power of Imitation.

The Hon. Daines Barrington has given a table
of the comparative merits of singing birds with
regard to their notes,—20 being the point of perfection: In which the nightingale's is said to
be 19 in mellowness, 19 in plaintiveness, 19 in
compass, 19 in duration, and 14 in sprightliness.
He also says, that the sound of its song filled
" the circle of an English mile;" and that, when
it sang round the whole compass of its warble, he
remarked sixteen different beginnings and closes.
The notes between were varied with so much
seeming art, and the manner of modulating them
was so skilfully managed, that the effect was both
grand and pleasing. Kercher and Barrington both
attempted to note this bird's song in musical types;
but, although the notes were played by an excellent performer on the German flute, they bore no
resemblance to the native warble of the bird;
owing, as Mr Barrington conjectured, to the im-
possibility of marking the musical intervals; for
the measure was so varied, the transitions so in-

sensibly blended, and the succession of tones so
wild and irregular, as to soar far beyond the fetters
of method.

The nightingale's song is finely described by the
ingenious author of *l'Histoire des Oiseaux*. He
says :—" The leader of the vernal chorus begins
with a low and timid voice, and he prepares for
the hymn to nature by essaying his powers and
attuning his organs; by degrees, the sound opens
and swells—it bursts with loud and vivid flashes
—it pants and murmurs—it flows with smooth
volubility—thus pouring out the soft breathings
of love and joy."

The female of this bird sings sweetly, but not
powerfully. Virgil and Milton have been criti-
cised for attributing to her this property, said to
belong exclusively to the other sex. In this the
poets were correct; but the ignorant will always
condemn, as absurd or as false, what their intellect
cannot comprehend.

Nightingales may also be taught to articulate
words; and, though we may not believe *all* that
Pliny says, and even think that the Ratisbonne
story, mentioned by Gesner's friend, was perhaps
a dream, we shall not assert that it is impossible
for two of these birds to hold a conversation with

each other, or to pick up and rehearse what they may hear: but we shall leave that query to be decided by those who are better acquainted with the wonderful powers of these birds than we are. This we do know, that they may be taught to imitate the airs of a nightingale's pipe, and to chaunt a waltz, or to take a part in a duet, and keep in time and tune. By proper tuition, kindness, and perseverance, it is really astonishing what they may be taught, their ear is so excellent and their voice both powerful and flexible: but the finest airs, chaunted in the purest style, have no charms for us, when compared with their own sweet, thrilling, plaintive, wild notes.

The Italians and the French are good teachers of these birds; but the Germans are the best.

Nightingales, when once domesticated, get very tame, and become much attached to those who feed them. The step of their benefactor is to them a pleasing sound, and they welcome it with notes of joy; and, when deprived of those they love, they have been known to pine and die. " One that was presented to a gentleman, no longer seeing the lady who used to feed him, became sullen, refused to eat, and was soon reduced to that state of weakness that he could no longer support himself on his perch; but, on being restored to his former mistress, he quickly recovered, ate, drank, returned to his perch, and was well in twenty-four hours. Buffon mentions one that lived to the age of seventeen years; and, though hoary, was yet happy and gay, warbling as in early youth, and caressing, to the last, the hand that fed it."

It is said that nightingales are neither to be found in the western counties of England, nor in Wales, nor farther north than Yorkshire, and certainly not in Scotland. We have been told, however, by a gentleman, that these birds have been heard in Dumfriesshire, and in Ayrshire; and a circumstance, mentioned to us by another gentleman, leads us to believe that nightingales do occasionally visit Scotland: they may possibly be stray birds, or birds, having lost their mates, in quest of others. But the circumstance mentioned by our friend was this:—that, in June, last summer, 1822, in his garden, near Leith Walk, Edinburgh, he heard a bird sing every evening, for about ten days, until a late hour, sometimes till near midnight. The song was quite new to him; but it was much sweeter, and continued longer in one strain, or warble, than the song of any other bird with which he was acquainted. He describes it thus: At first it was loud but mellow; then sprightly and varied, the cadence dying away as if the bird was more distant at the close of the warble than when it first began; but, the next moment, he heard it as before, thus showing the bird was still on the same tree. From his description of this bird's singing, we are almost led to conclude it really must have been a nightingale. At this time a pair of magpies returned to their old nest, which was placed on a large tree near his house, and, from that time, the bird was no longer heard. Whether nightingales have ever bred in Scotland, we cannot say; but if such an expedient as the following was attempted, we think it might succeed: Were birds procured and reared from the nest, and set at liberty the next season at pairing time, in some sequestered, sheltered spot where there was covert, *viz.* hedges and brushwood, or groves and plantations through which a clear rivulet ran over sand-banks, and pebbles,—and many such situations are to be found in Scotland,)—it is possible they might breed, and, after that, there would be no danger of losing them, as it is well ascertained that migratory birds have a strong inclination to return to and nestle in the places where they themselves were bred; and thus this delightful warbler might be added to the song-birds of Scotland; although we have been told Scotland already possesses a superior warbler to this bird. Such at least was the opinion of an Ayrshire laird, who, after hearing a nightingale in the south-east of England, thus answered a gentleman who asked him if any songster in Scotland could compare with the bird he had just heard:—" Div ye ca' that singing? Did ye ever hear a whaup?" " No." " Weel, I wadna gie the wheefle o' a whaup* for the sang o' the best nightingale in England."

* The whimbrel.

The gale brings music! and, in strains of sorrow,
Stealing along the glen, so sad, so sweet,
Some elfin harp is surely strung to sadness;
Or 'tis the requiem chaunt of holy men,
For knight or palmer gone,—or wailing dirge
Of some lone minstrel, for the brave and good.
Tho' mournful is the strain, no mourner sings:
It is the vesper hymn of some sweet bird
Chaunting his evening lay to yon bright star,
The while, the plaintive cadence soothes his mate.

Anonymous.

THE BLACKCAP.

MOCK NIGHTINGALE, OR NETTLE CREEPER.

SYLVIA ATRICAPILLA; LATHAM.—MOTOCILLA AT-
RICAPILLA; LINNÆUS.—LA FAUVETTE A TETE
NOIRE; BUFFON.

THE Blackcap is truly a most delightful warbler, and may be ranked as second in the class of

British song-birds. Indeed, in our opinion, its
mellow notes are equal if not superior in richness
of tone to any in the nightingale's song. It is true
the warble is desultory, but sweetly wild and full
of melody. The cadence rises and swells, then dies
away in a soft and plaintive strain. Its shake or
trilling note is the finest we ever heard : A first-
rate opera-singer might imitate it; but, like all
imitations, it would fall short of the original.
This bird is not very common; but it appears to
be not unfrequent in the vicinity of Edinburgh.
We have heard it at Duddingston, the Dean, Ra-
velstone, and Craig-crook, all within about two
miles of the above mentioned city. It haunts
shrubberies and young plantations, but generally
near some old house or castle; also orchards and
gardens.

Montagu says " The blackcap is a migratory
species, visiting us early in the spring, and retiring in September." It "frequents woods and thick
hedges, and seems particularly partial to orchards
and gardens, where it delights us with its charming melodious song, which is very little inferior
to that of the nightingale's, except in variety of
notes."

The cocks arrive in Britain about the middle

of April, and the hens in ten or twelve day
On the first arrival of these birds they feed
berries of ivy, hawthorn, spurge laurel, a
vet, which is also the food of the few that
with us during the winter: Insects, howev
their favourite dainties, and with them th
their young, such as soft grubs, butterflie:
and ant-eggs. The male is very assiduo
attentive to the female during incubation.—
gularly takes his turn of sitting on the eg
about nine or ten in the morning till four
o'clock in the afternoon, and, while she is
he occasionally supplies her with food, and
her in the tenderest manner with his
When the young leave the nest, which
early if any way disturbed, they then foll
parents, hopping from spray to spray,
wards evening the whole family take po
of a branch to roost on for the night, the n
in the middle, and the parents at each en
ing close to their progeny to keep them v
A helpless family and tender parents so a

becomes very tame. It is also of an affectionate nature, and gets very fond of its master or mistress. On their approach, it flutters towards them, and welcomes their visit with a peculiar note. From the warmth of its affections, Mademoiselle Des Cartes could not forbear remarking that, with all deference to her uncle's opinion, it was endowed with sentiment.

.. More might be said of this charming warbler; but we shall conclude this part by observing that the young and old birds, with respect to food, general treatment, and diseases, are managed in the same manner as the nightingale and redbreast.

The cock is easily known from the hen, by the black on his head, from which these birds get their name. The head of the hen is of a light reddish-brown, and she is altogether paler in colour.

Of the Nest and Eggs.

These birds build in May, and the young should be taken in June. The nest is generally placed in a bush or fir tree near the ground, sometimes in a bush of evergreen, eglantine, or woodbine; but

their favourite spots for nestling are shrubberies
and young plantations, under the lower branches
of larch, spruce, or silver fir. That described by
Pennant was placed in a spruce fir about two feet
from the ground: the outside was composed of
dried stalks of the goose-grass, with a little wool
and green moss round the edge; and the lining
consisted of fibrous roots thinly covered with
horse hair. It contained five eggs of a pale reddish-
brown, mottled with a deeper colour, and sprinkled
with a few dark spots. According to our nomenclature their colour is very pale chesnut-brown,
inclining to flesh-red, freckled with deep chesnut-
brown, and marked at the large end with larger
spots of liver-brown.

Description and Plumage.

The blackcap is slender and elegant in form,—
what the Scotch would express by the word
" trig." It is quick and lively in its motions, and
seems to turn round the trunks and branches of
trees to hide itself as it were from observation,
like some of the wrens and fauvettes. It is about
five and a half inches in length: the upper man-

dible is yellowish-brown, the under one pale bluish-grey; eyes hazel or umber-brown; crown of the head of the male bluish-black, breast pale but dingy ash-grey, slightly passing into pale wood-brown; back pale clove-brown passing into yellowish-grey; towards the tail, inclining to dull white; the wings and tail the same as the back, but darker and still more dingy; the edges of the feathers of both faintly inclining to a very dingy pistachio-green; legs and feet bluish-grey. The hen is rather larger then the cock-bird, and is easily distinguished from him by the crown of the head, which in her is of a dull chesnut-brown; she is also paler, and altogether more of an ash-grey colour.

Song.

The first time we heard the notes of this bird, we mistook them for those of the redbreast and the thrush, such was the similarity of its notes to to theirs, particularly its low notes, which are so similar to the soft, mellow tones of the thrush, that, when the latter are heard from a little distance, it is not easy to distinguish them from those of the blackcap; but on hearing it several times, and at last seeing the bird, and observing the motion of its little throat, we were convinced the whole notes proceeded from one bird. On mature consideration, (having now heard it frequently,) we are still of the same opinion; but when it pours out its full song, we think it possesses many notes very similar to several in the nightingale's warble, thus combining in its song many of the musical qualities of these three excellent song-birds. Buffon says, " that its airs are light and easy, and consist of a succession of modulations of small compass, but sweet, flexible, and blended." And our ingenious countryman, Mr White, observes:—" that it has usually a full, sweet, deep, loud, and wild pipe; yet the strain is of short continuance, and its motions desultory. But when this bird sets calmly, and in earnest engages in song, it pours forth very sweet but inward melody, and expresses great variety of sweet and gentle modulations, superior, perhaps, to any of our warblers, the nightingale excepted."

Some redbreasts love amid the deepest groves,
  Retired, to pass the summer days: Their song
Among the birchen boughs, with sweetest fall
Is warbled, pausing,—then resumed more sweet,
More sad, that, to an ear grown fanciful,
The babes, the wood, the men rise in review,
And robin still repeats the tragic line.
                    Grahame's Birds of Scotland.

THE REDBREAST.

ROBIN REDBREAST, OR RUDDOCK.

SYLVIA RUBICULA; LATHAM.—MOTACILLA RUBI-
  CULA; LINNÆUS.—ROUGE GORGE; BUFFON.

THIS delightful little warbler, equally sacred to
the cottager's hearth, the farmer's hall, and the
squire's mansion, is well known through the popular and piteous story of " *the children in the
wood.*" Its confidence in man has rendered the
redbreast a general favourite; and its familiarity
has procured for it, in most countries, a peculiar

R. Scott Sculp.ᵗ

ROBIN-REDBREAST.

name, such as might be given to some welcome
annual visitor: With us it is called Robin Redbreast;—in Germany, Thomas Gierdit;—in Norway, Peter Ronsmad;—and in Sweden, Tomi
Liden.

, The plumage of the redbreast, though harmonious, is plain; and it is rather remarkable that all
our finest songsters have but few showy colours.
Though the redbreast is so well known to man, yet
naturalists are still doubtful whether to consider
it as a migratory or stationary bird. Buffon says,
that it migrates singly, not in flocks: Many, however, remain with us through the winter; but these
appear, (at least such is our opinion,) to be all males.
During severe storms, when the ground is covered
with snow, this bird approaches the habitation of
man with a confidence and winning familiarity
which always insure to the tiny stranger kindness
and protection. He has been known to come to a
window,—to tap, and, if it be opened, to enter, to
eye the family in a sly manner, and, if not disturb
ed, to approach the board, pick up crums, hop
round the table, and catch flies if any remain, then
perch on a chair or window-cornice; and, finding
his situation comfortable, is often seen in this
familiar way to introduce himself to the family,

and to repay, with seeming grati
tality, by the melodious warbl
throat: and this daily throughou

We know a gentleman who, last
caught a young redbreast, one
flown in his garden: A short tin
was lost, several days elapsed, a
appear; when the gentleman, w:
den with a friend, saw a bird of t
he thought very like his, hop;
or five others, that seemed to b
age. He requested his friend
returned to the house for a few
held in his hand, and callin;
bird appeared to recognise the
accustomed to, perched upon h
instantly secured. The bird is r
full plumage, and singing deligl
at liberty through the room, fo
large, light, and airy cage, the do
open, he seldom enters it. In t

a seed of hemp, and calls " *Robie!*" he
flies at it, picks it from between the
thumb, darts off, and this so rapidly
cannot detect how he extracts the seed.
fine healthy bird, in full feather, thoug]
on hempseed,* loaf-bread, and what 'fli·
catch, with now and then a spider.

His manner of feeding is rather curious
of bread is put down, which he pecks a1
point, generally near the centre of the p
he has made a hole through it; he then
another place, and does the same: He i
quisitive, and it is amusing to observe
any thing is brought into the apartmen
books, paper, &c.—At first he advances
caution; but, finding the object motionle
tures nearer, hops round it, but neve
content till he has got upon it, and nev·

* About six weeks since, and after this article wa
bird lost his voice; his eyes watered much and app·
held up his head, seemed dull, drooped, and was
well.—the effect, perhaps, of too much hemp-seed.

unless disturbed, until he has examined it with the eye of a curious inquirer.

One morning, a roll of paper, more than two feet long, being laid on the table, Robie instantly saw it was a new object, flew to it, hopped round and round it several times, and at last finding it impossible to satisfy himself without a narrower inspection, he hopped in at the one end and out at the other.

We have heard many anecdotes of the redbreast, but what we have mentioned will suffice to show its manners in a state of domestication. This bird may be taught various pretty tricks, and even to articulate words. We know that a lady in Edinburgh possesses one who very distinctly pronounces, " How do ye do?" and several other words. Her method was, early in the morning, before giving it any food, to repeat very often what she wished it to learn.

In a wild state, these birds are very pugnacious. Each cock seems to have certain bounds, which he considers his own, and within which he will allow no other bird of the same species to range. The redbreast builds its nest in different situations according to circumstances: We found one at the edge of a rocky bank near Roslin, but so hid by grass and ivy, that, had it not been for some wild flowers for which we were looking, (the hen sits so very close,) we might never have found it. Last year, (1822,) at Craig-Lockhart, near Edinburgh, we saw a cock-bird rather agitated, with something in its bill, and, thinking the nest might be near, we were anxious to see if they built in so exposed a situation as the way-side: After much trouble, and careful examination of both sides of the road, we at last discovered it by the hen flying out, when we were within a foot of the nest; had she not been on, it was so curiously concealed, we might never have perceived it.

In a garden at Canonmills, for several years, a redbreast, (we believe the same bird,) has built its nest; once in a bower, another time in a laurel close by a wall, and last year artfully hid amongst ivy on the trunk of an old willow-tree: it was found by observing the cock going in with food, and, just as our hand was at the hole which led to the nest, the bird flew boldly down from a tree, and struck at our fingers.

This winter, (1822–3,) the same redbreast watched when the servant went at dusk to shut up a greenhouse in the garden, entered with her, and coming near, pecked the crums which she held to it from her hand,—remained all night, and was ready in the morning for the same fare. When she returned to open the door, he usually came out with her, (unless in very bad weather,) and flew to the garden, and, as she repaired to the house, poured forth a strain of grateful melody; and this he did regularly almost every day during this very severe winter.

Of the Nest and Eggs.

The nest is composed of bent, dead leaves, grass-roots, and other fibrous substances, mixed with moss, and lined with thistle-down, hair, and feathers. The eggs, four or five in number, are of an orange-coloured white, freckled, particularly at the large end, with pale orange-red spots, inclining to brown.

To find the Nest.

It is desirable to know how to look for the nest, it being of consequence to get the birds young, if we wish to tame, or teach them any pretty tricks. When you see a redbreast, observe if it has any thing in its bill: Do not frighten it, and it will soon go to the nest; but its instinct is so great that it sometimes flits about before entering the nest: wait therefore until it has gone in and out several times from the same place; when in, steal upon it quickly, otherwise the opportunity may be lost of scaring it, this being the best method of discovering the nest; for, if you do not see the very spot from whence the bird springs, its mossy mansion is so artfully concealed, you may not, after all, be able to find it. The same rule holds for taking the nests of nightingales, wrens, blackcaps, and most of the soft-billed species, which, being the shiest birds, display the greatest ingenuity in concealing their nests. The redbreast builds in April, May, and June, and has sometimes two broods in the year. The young birds may be taken from eleven to thirteen days old, at which time approach with great caution, laying the hand upon the nest, and don't breathe on the young till secured, lest they scramble over the nest, and the whole be lost.

Treatment of the Young.

Put the nest in a clean, dry cage, and, while it (the nest) continues clean, let the young remain in it till they gather strength and incline to perch.

Keep them warm, dry, and very clean; put dry moss round the nest and on the bottom of the cage; for the cold wood, while they are so young, is apt to give them the cramp. Feed them with butcher-meat, very finely minced, but free from fat, mixed with crums of loaf-bread rather old, but sweet. The least sourness or bad yeast in the bread will kill them. New bread is also detrimental. Add to the butcher-meat and bread some bruised hempseed and a little of the yolk of an egg very hard-boiled; mix the whole together with water, so as to make a thick paste; feed them with this every two or three hours, beginning as early as possible in the morning, particularly at first, giving them occasionally a spider or meal-maggot, (the *larva* of *tenebria molitor*,) which may be had at meal-dealer or at baker shops. Nothing more is necessary till the birds take the perch; the nest and moss must then be removed, and the bottom of the cage sprinkled with fine dry earth from an ant-hill, mixed with a little fine gravel, and a few ant-eggs put on the gravel will be beneficial. The mixture already mentioned must be put in a clean flat dish on the bottom of the cage, and in another dish a little pure water. Change both daily, as the health of the birds depends on fresh food and wa-

ter, and being kept dry and clean. The same
food and treatment answers for adult birds. The
young redbreasts are very unlike the parents, being spotted with dark-brown on a lighter-brown
ground, and having a cry similar to that of the
young hedge-sparrow.

. The manner of taking old birds by trap, or
limed twig, and their after treatment, is the same
as practised regarding the nightingale.

Diseases.

Damp and cold are detrimental to redbreasts,
and may produce cramp. To prevent it, keep the
birds warm, clean, and dry. And to cure it, give
them two or three meal-worms every day. For
gout, rub their legs and feet with fresh lard or
sweet butter, and occasionally give them a few meal-
worms. For giddiness, give them a few spiders,
or two or three hog-lice, in Scotland called sklaters. When moulting they are subject to the pip,
(*Scottice*, " roop,") a swelling near the rump: this
must be opened with a needle, and then anointed
with lard. If neglected, they will mope, perhaps die.
. The gapes, that is, when they open the bill,

pant, and appear as if choking, and like to suffo-
cate, this, if not attended to, may occasion death.
In this case, either the nostrils are stuffed, or it
proceeds from animalcule in the larynx or throat;
if in the nostrils, take a fine-pointed needle and
pick out what seems to obstruct their breathing;*
but should the disease still remain, then it is vermin in the windpipe. For this put a little saffron
or liquorice-root infused in the water, to which
add a little white sugar-candy. Should they bunch
up their feathers, or appear dull, or put their bills
under their wings, give them a spider or an ear-
wig two or three times a day, or a bruised hemp-
seed, of which they are very fond; when well, they
will swallow it whole. Saffron or liquorice may
also be of service in bringing them into spirits a-
gain, and in improving their song. When their
claws are too long, they must be paired, otherwise
they get entangled on the perch: this frightens,
and makes them pant, which is apt to produce fits.
All small birds are liable to these diseases, and
the treatment is in most cases the same.

* We omitted to notice, under the articles *Blackbird, Thrush,*
&c. that their nostrils get clogged if the food is too soft; in this
case they breathe with difficulty. To cure it, take a small feather
and draw it quite through the nostrils, which is easily done, as they
are pervious, and in a few minutes the birds will be quite well.

Description and Plumage.

The redbreast is between five and six inches in length, bill slender, and of a horn-colour. Eye black, large, full, and mild, with a small orange-red circle round it. Upper parts of the plumage, *viz.* head, back, and wings, pale umber-brown inclining to olive, in some lights appearing as if tinged with yellowish-brown. Forehead, throat, and breast, rich orange-red; lower parts, greyish-white, with a mixture of dull cream-yellow; legs, clove-brown, passing into umber-brown; claws black.

The hen is very like the cock; but neither so large or full of spirit, nor so bright in the plumage. To choose a cock-bird, let him be large and sprightly, having a full sparkling eye; the brown on the back, rich, glossy, and dark, and the red on the breast large and bright; this last is the best criterion to judge by.

Song.

The redbreast will learn the notes of other birds; but his own being so fine, it is a pity to spoil it by teaching him to imitate other warblers: His song is rich, full, melodious, melting, and tender; it is very various, at one time having a deep meloncholy tone, broken with sprightly turns between; then mellow and plaintive. The spring and autumnal notes are different: In spring his melody is rich, but quick, softly-melting, and dying away in harmonious cadences; in autumn they are plaintive, but still more rich and sweet,—as if he sung the dirge of summer, or wailed the departing year.

R. Scott Sculp.

REDSTART.

Published by John Anderson 'Jun' 55, North Bridge Street, Edinburgh. 1823.

THE REDSTART.

SYLVIA PHÆNICURUS; LATHAM.—MOTACILLA PHÆ-
NICURUS; LINNÆUS.—LE ROSSIGNOL DU MU-
RAILLE; BUFFON.

THE plumage of this bird is beautiful, and its
shape elegant: Its motions are lively and quick,
and, though inferior in song to the three preceding
birds, yet, from the beauty of its plumage and the
sweetness of its warble, it is considered, by many,
a desirable bird for the cage: it is a migratory
bird, and, in many parts of this island, rather a
rare species.

It visits us in April, and departs in September.
Its haunts are local. Montagu says it is rarely
found in Cornwall, and perhaps not farther west
than Exeter in Devonshire; and another author
observes, " that it is seldom seen farther to the

north than Yorkshire." But we have frequently met with it in the neighbourhood of Edinburgh. Though a very shy bird, it often approaches and builds near the habitations of men, and constructs its nest in places that we would scarcely expect so timid a bird would select for that purpose. At Craigcrook-Castle, near Edinburgh, we found its nest in a hole of a wall close by an old gateway, through which people daily pass to the castle; it was placed within reach of the hand from the ground. These birds often haunt orchards, gardens, and shrubberies; but they also frequent solitary situations among rocks, crags, and woods, where they build in the crevices of dangerous ravines and precipices. Though wild and timorous birds, they are often found in cities, but always selecting the most difficult and most inaccessible places for the important work of incubation. If the eggs are touched by the hand, unless the hen has sat some time, she will forsake the nest and build again.

Redstarts feed on insects, their larvæ and eggs, also on wild berries. The old birds are useless for the cage, as they pine and die; but the young are easily tamed, and are reared and managed in every respect as young nightingales. The redstart,

when it alights, has a curious vibratory motion
with its tail.

Of the Nest and Eggs.

These birds in their manners are very much
like titmice, and they build in similar situations;
at least those nests we have found were in holes
of old garden and orchard-walls covered with
ivy, or in the holes of trees near such places.

The outside of the nest is formed of moss; the
inside is lined with down, feathers, and hair; and
the eggs, from four to five in number, are not unlike those of the hedge-sparrow, but rather smaller,
longer in proportion, and a little paler in colour.
They are of a pale tint composed of verditer-blue,
passing into verdigris-green.

Song.

Its natural warble is neither vary varied, loud,
nor extensive; but its notes are sweet, and the modulations are finely blended and tender. When
young, and placed by good song-birds, it soon acquires another strain, and thus becomes an excellent bird for the cage.

Description and Plumage.

The length of this bird is between five and six inches. It is of an elegant slender make, the bill and eyes brownish-black or dark hazel,—forehead snow-white,—part of the neck, the checks, and throat, bluish-black, the black on the throat extending a little down on the breast,—head, neck, and part of the back, beautiful ash-grey,—breast and rump, fine reddish-orange, lower parts a whitish-yellow inclining to yellowish-grey,—tail, reddish-orange, excepting the two middle feathers, which incline to chesnut-brown,—legs, feet, and claws bluish-black. The female is yellowish-brown, inclining to grey where the male is black, excepting the chin which is whitish; and where he is red, she is more of a pale but dingy chesnut-brown. The young are spotted all over, somewhat like the young of the redbreast.

That strain again ; it had a dying fall !
Oh it came o'er mine ear like the sweet south,
That breathes upon a bank of violets,
Stealing and giving odours.

Twelfth Night.

GREATER PETTYCHAPS.

SYLVIA HORTENSIS; LATHAM.—MOTACILLA HOR-
TENSIS; LINNÆUS.—LA FAUVETTE; BUFFON.

Sir Ashton Lever was the first who noticed this
bird as a British species, and sent it to Dr Latham. It is rather a rare bird even in England, and
still more so in Scotland. We first heard its warble on Corstorphine-Hill, about two miles to the
westward of Edinburgh. We have since heard it
in Roslin woods.—We were not aware what bird it
was till we read Montagu and Bewick, and, from

their description, we conclude it to be the same.
As it is a wild, shy, and timid bird, it was with
considerable difficulty we obtained a sight of it.
We heard it first among some low bushes, afterwards in a sloe-bush, but, on our approaching
nearer, the warble ceased; however, on our wait-
ing a little, we again heard the strain, and, looking
over a steep bank, where we thought the warbler
was, perceived it perched on the topmost branch
of a tree below us. We knew it to be the bird from
which the delightful melody came, both by the
direction of the sound and by the motion of its
little throat. It appeared to us much like the
hen blackcap, but darker and more of a brownish-
green colour. Though its notes are different from
those of the nightingale's, yet, in our opinion, it
falls little short of that bird as a songster. We never
had the nest of the pettychaps, and never saw the
bird nearer than what we have already stated.
We shall therefore close the account of this most delightful song-bird, with what that eminent ornithologist, the late Colonel Montagu, and Mr Bewick,
say of it. Colonel Montagu remarks :—" Its song
is little inferior to that of the nightingale. Some
of the notes are sweetly and softly drawn; others

quick, lively, loud, and piercing, reaching the distant ear with pleasing harmony, something like the whistle of the blackbird, but in a more hurried cadence: Sings frequently after sunset. This bird chiefly inhabits thick hedges, where it makes a nest composed of goose-grass and other fibrous plants, flimsily put together, like that of the common white-throat, with the addition of a little green moss externally: the nest is placed in some bush near the ground. It lays four eggs, about the size of a hedge-sparrow's, weighing about thirty-six grains, of a dirty white, blotched all over with light-brown, most numerous at the large end, where spots of ash-colour also appear. In Wiltshire, where we have found this species not uncommon, it resorts to gardens in the latter end of summer, together with the whitethroat and blackcap, for the sake of currants and other fruit."— And Mr Bewick says :—" This bird frequents thickets, and is seldom to be seen out of covert; it secretes itself in the thickest parts of bushes, from whence it may be heard, but seldom seen. It is truly a mocking-bird, imitating the notes of various kinds, generally beginning with those of the swallow, and ending with the full song of the black-

bird : we have often watched with the utmost attention, whilst it was singing delightfully in the
midst of a bush close at hand, but have seldom
been able to obtain a sight of it."—As this is a soft-
billed bird, we think the young and the old birds
ought to be fed and managed in the same manner
as the nightingale, &c.

THE HEDGE WARBLER,

HEDGE-SPARROW, WINTER FAUVETTE, OR FIELD-SPARROW.

SYLVIA MODULARIS; LATHAM.——MOTACILLA MO-
DULARIS; LINNÆUS.——LA FAUVETTE D'HIVER;
BUFFON.

THOUGH this bird has no great variety of notes
in its natural song, yet those it has are rather
sweet; and, as it sings so early as the beginning of
the year, if the weather be fine, and from its
quickness in acquiring the notes of other birds,
we have given it a place among the song-birds.
In a state of confinement it soon grows very familiar, and becomes very much attached to those
who feed it. If placed beside good song-birds, it
speedily picks up their notes, and in a short time
becomes a soft, sweet, agreeable warbler; it may
also be easily taught to pipe. This bird is so com-

mon, and so well known, that we need give no description of it. The young are spotted, hop about, and chirp so like young redbreasts, that it is not easy, in that early state, to distinguish the young of the hedge-sparrow from those of the redbreast.

The winter fauvette is seldom found in wild districts, but seems to prefer nestling near the habitations of men. It may be seen both in winter and summer about the roots of hedges, (from whence it derives its common name,) and among bushes in gardens and fields. In summer it feeds on insects; and in winter comes near our doors, to pick up crums of bread, seed, &c. It builds early in spring, commonly in hedges or low bushes; the nest is made of moss and wool, well lined with horse-hair. The eggs, from four to five in number, are of a clear, beautiful blue colour, of a tint between pale verditer blue and verdigris green. It is in the nest of this bird that the female cuckoo drops her egg, which the hedge-sparrow hatches, and also brings up the young cuckoo, which last, by a singular process, throws out the eggs and young of the hedge-sparrow, and takes possession of the nest; and it is curious and amusing to see the young cuckoo following so diminutive a stepmother to be fed. The whole is certainly a phenomenon in natural history.

We have frequently scared the hedge-sparrow from her nest; when, to draw us away from the spot, she appeared as if her leg or wing was broken. And " if a cat or other voracious animal should happen to come near the nest, the mother endeavours to divert it from the spot by a stratagem similar to that by which the partridge misleads the dog,—she springs up, flutters from place to place, and by that means allures her enemy to a safe distance."

Young and old hedge-sparrows are managed fed, and treated, in every respect, like young and old redbreasts.

G

SEDGE-WARBLER,

OR REED-FAUVETTE.

SYLVIA SALICARIA; LATHAM.—MOTACILLA SALI-
CARIA; LINNÆUS.—LA FAUVETTE DE ROSEAU;
BUFFON.

THIS bird is of an elegant form, and the colours of the plumage are very harmonious and
finely blended. Its natural song, though weak, is
sweet, and rather varied. At times it will imitate the warble of the skylark, then the twitter of
the swallow, and even the chirp of the house-
sparrow. It is, therefore, ranked among the English mocking-birds.

Its haunts are fens and swampy places, where
bushes, reeds, and sedges grow; and there it
builds its nest, which is placed either amongst
rushes, or ingeniously fastened to three or four
reeds; and in this floating cradle, though rocked
by the tempest, the hen securely sits, without fear
or dread. The nest is composed of coarse grass

and small stalks of reeds, and lined with fine grass
and hair. The eggs, from four to six in number, are
of a pale, dingy, yellowish-grey, inclining to hair-
brown, and mottled with hair and broccoli-brown
spots. Though this is a wild, timid, and shy bird,
we have had its nest not a gunshot from the vil-
lage of Bell's Mills, about a mile from Edinburgh.
Mantagu says—"The song of this bird has been er-
roneously given to the reed-bunting by various
authors, whereas that bird has no notes that de-
serve the name of song." But, with deference to
that accurate ornithologist, we know the reed-
bunting sings very sweetly, though soft and low;
not unlike some of the low notes of the grey linnet. Montagu adds:—" While this little warbler,
concealed in the thickest part of a bush, is heard
aloud, the bunting is perched on the upper branches,
and therefore thought to be the song-bird; and
thus their songs have been confounded." But we
have frequently heard and seen the reed-bunting
in low hedges, where we have had their nests, and
where we never either heard or saw a sedge-warbler; besides, their notes are very unlike each
other. But it is quite correct, that, if it (the sedge-
warbler,) is silent, a stone thrown into the bush
will set it a-singing instantly.

It feeds on dragon-flies, may-flies, or ephemeræ,

and other insects that frequent marshes; and it is curious to see how quickly it darts from the reeds or willow-roots, catches the fly, and flits back again. This we have often observed. The reed-warbler sings a great deal, both through the day and in the evening, but never in sight of any person, if they are in motion. In order to see the bird, it is necessary to sit or to lie down. These birds may be brought up and fed as the redbreast or hedge-sparrow.

Description and Plumage.

The reed-warbler is an elegant species, both as to shape and plumage; length rather less than five inches and a half. The bill hair-brown above, whitish below; eyes hazel, or umber-brown; head, neck, back, wings, and tail, yellowish brown, faintly inclining to oil-green; middle of the feathers of the back of the neck and wing-coverts marked with a dingy, very pale umber-brown tint; over each eye there is a streak of yellowish-white; throat the same, mottled with pale hair-brown; breast and lower parts a beautiful pale primrose-yellow, inclining to white, the breast rather deeper, passing into wine-yellow; tail rather short, and a little rounded at the end; legs and feet hair-brown; hind claw somewhat long, and a little curved.

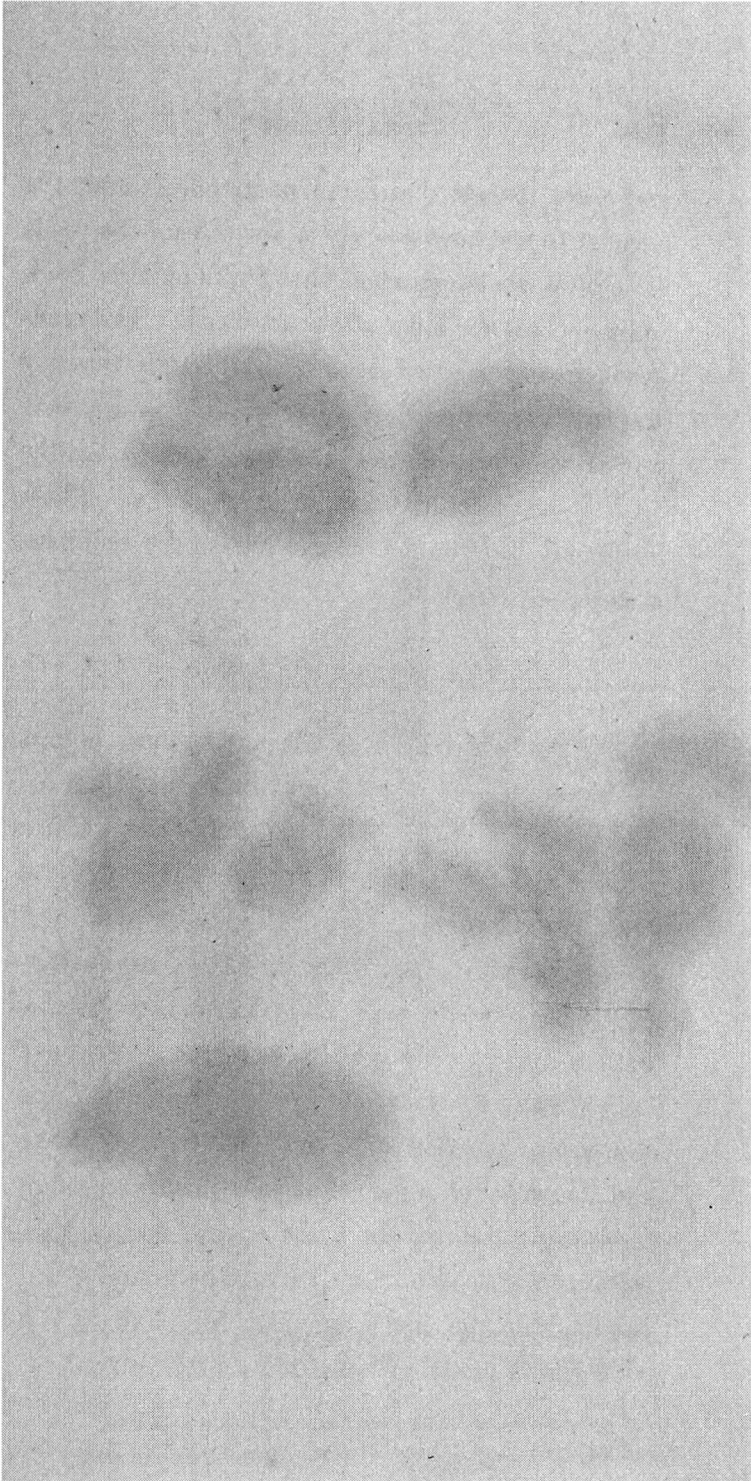

THE WREN.

Published by John Anderson Jun.ʳ 55. North Bridge Street, Edinburgh. 1823.

THE WILLOW-WREN.

GROUND-WREN, OR YELLOW WILLOW-WREN.

SYLVIA TROCHILUS; LATHAM.—MOTACILLA TRO-
CHILUS; LINNÆUS.—LE POUILLOT, LE CHANTRE;
BUFFON.

THERE are three wrens so similar in size, plumage, and habits, that they have been often confounded:—the yellow-wren of Montagu, or yellow
willow-wren of White and Bewick; the willow-
wren of White and Bewick, or the reed-wren of
Montagu; and the wood-wren of that author, or
the least willow-wren of White and Bewick. We
shall confine ourselves at present to the willow, or
yellow willow-wren; because, when we describe it,
we nearly describe the other two,—but still more,

because it is by far the finest warbler of the three: Its notes are somewhat like those of the redbreast, but not so loud, or so mellow, though wildly-sweet, and very plaintive. We have often seen the willow-wren shot when in the act of singing, and therefore cannot mistake the species. It " is rather larger than the other two, and in shape it is some-what similar to the hen blackcap, but its plumage is more of an olive-brown and yellow colour." It haunts trees, where we have often heard it pouring out its melodious song from the highest branches; but it builds on the ground, generally in banks, where we have often found its nest, but always in sequestered situations, amongst open brushwood. We have had one almost covered by primroses and their leaves, also among brambles, and at the root of a wild rose-bush; it conceals its nest with great ingenuity. Mr White of Silbourne, and a friend of his, observed a bird of this species as she sat on her nest, but forbore disturbing her: A few days after, as they were passing the spot, they looked for the nest, but could not discover it, until Mr White removed a tuft of green moss, which seemed thrown as if at random, over the little dwelling, so completely, as to conceal it from all intruders.

The nest is made of moss and grass, and lined with feathers and hair. The eggs, from five to seven in number, are white, sprinkled with pale brownish-orange spots. All the three species of willow-wrens are migratory: they arrive in Britain in April, and depart in September. If reared for the cage,—and this species is really worth the attention of amateurs,—they are fed and managed as the redbreast and other soft-billed birds.

Description and Plumage.

The willow-wren is very slender in its make, and long in proportion to its size. In length it is about five inches and a quarter; the bill hair-brown above, dingy straw-yellow below; eyes hazel, or umber-brown; over the eyes there is a pale-yellow streak; head, back, wings, and tail, very pale-yellowish brown, inclining to wine-yellow; breast and under parts pale, but dingy primrose-yellow, streaked on the breast with some darker marks of the same colour; wing-coverts the same; legs and feet wood-brown.

GOLDEN-CRESTED WREN.

SYLVIA REGULUS; LATHAM.—MOTACILLA REGULUS;
LINNÆUS;—LE ROITELET, LE SOUCI ; BUFFON.

This beautiful bird is perhaps the smallest in Europe : it is certainly the least in Britain.—Though its natural song is weak, the notes are sweet, and may be rendered still more so by placing the bird near other songsters. For this reason, and for the beauty of its plumage, we would recommend it to amateurs. The golden-crested wren is rather common, more so at least than is generally supposed. We have seen it in the woods on Corstorphine Hill, near Edinburgh, in the woods at Roslin, and other places in the neighbourhood. Though so small, it braves the severest winters of Britain, and even of countries situated in more northern latitudes : It haunts deep woods and sequestered forests, where it likes to build. Its manners are similar to those of the titmouse. We have seen it flitting from tree to tree, and creeping up their trunks, searching for insect food,—exactly like the titmice.

Montagu remarks:—"The nest is not made with an opening on one side as described by some, but is, in form and elegance, like that of the chaffinch, composed of green moss, woven with wool, and *invariably* lined with feathers, with which it is so well bedded as to conceal the eggs." This we know to be true; and, to corroborate the fact, a gentleman, on whose knowledge we can depend, found one concealed under the branch of a larch tree in an old avenue : it was lined with feathers and down, and contained twelve eggs of a dingy reddish-white colour, faintly inclining to brown, particularly at the large end; their size was not much above that of peas : The cock was shot flying from the nest, and the hen was found in the nest, taken, and brought away with it.

Colonel Montagu found a golden-crested wren's nest containing ten young, in a fir-tree in his garden : he took it when the nestlings were about six days old, and brought it to his study-window to see

if the old birds would feed them. This they did;
he then brought the nest into the apartment; and
" it is rather remarkable, that, although the female
seemed regardless of danger, from her affection to
the young, yet the male never once ventured
within the room." On the contrary, the female
would feed them at the table, or while he held the
nest in his hand. The visits of the female were
generally repeated in a minute and a half, or two
minutes, or about thirty-six times in an hour,
and that during sixteen hours every day. Each
young bird weighed about seventy-six grains; and
they ate, in four days, nearly their own weight of
food. These beautiful little birds may be reared
and managed as nightingales and redbreasts.

Description and Plumage.

The golden-crested wren is about three and a
half inches in length; bill bluish-black; eyes, umber-brown, or hazel-colour. The crown of the head
is uncommonly beautiful: On it the feathers form
a crest; above each eye there is a line of velvet-
black; and, between the black lines, the feathers
are of a beautiful light Dutch orange-colour; the
forehead, chin, and cheeks, are of a pale-yellowish

grey, inclining to white; the neck, back, wings, and tail, are of an oil green colour, tinged with wax-yellow,—in some places faintly passing into yellowish brown, particularly the quills and tail, which are edged with wax-yellow : At the base of the secondary quills is a bar of bluish-black, above which the wing-coverts are tipped with yellowish white; the larger and smaller coverts are bluish-black, also tipped with yellowish-white; the breast and under parts are very pale wine-yellow, passing into yellowish-grey, terminating, near the tail, in yellowish-white; the tail is a little forked; legs and feet yellowish brown.

The male and female are very similar; only her crest is less, and of a gamboge yellow instead of a Dutch orange-colour.

Sweet warbler of the circling year,
Of summer bright and winter drear,
I love thy cheerful note to hear,
    Wand'ring the brakes among.

At early dawn thy native lay
Precedes the orient beam of day,
And oft, at evening's parting ray,
    I hear thy vesper song.

Within thy warm and mossy cell,
Where scarce 'twould seem thyself could dwell,
Twice eight, a speckled brood we tell,
    Nestling beneath thy wing!

And still unwearied, many a day,
Thy little partner loves to stay,
Perched on some trembling limber spray,
    Beside his mate to sing.

Anonymous.

THE COMMON WREN.

KITTY WREN.

SYLVIA TROGLODYTES; LATHAM.——MOTACILLA
TROGLODYTES; LINNÆUS.——LE TROGLODYTE;
BUFFON.

This little warbler is a very common species
in Britain. It remains with us the whole year, and
braves our severest winters. It begins its song early

in the morning, nor does it cease its warble till
late in the evening. It sings most part of the year,
even in winter; and sometimes, during a fall of
snow, it pours out its little roundelay. Some of
these birds retire to sequestered places to build;
others keep near the habitation of man, where
they may be seen flitting among the roots of
hazel, hawthorn, and tangled brushwood, near mills
and cottages; they also haunt sandy banks by the
sides of rivulets and mill-streams. Their principal
food is insects; but in winter, like the redbreast,
they approach cottages to pick up crums, &c.
They also pick bones that have been thrown out
from houses; and it is curious to see one of these
birds busy in the orbit of a sheep's skull. We have
also observed a blue titmouse do the same thing.
The wren sometimes builds in fir-trees, but often
under banks amongst the roots of trees, or in the
thatch of old cottages, stables, and other out-
houses; but, what is curious in the manners of
this little bird, instinct directs it to adapt the materials of its nest to the place it has chosen for
the purposes of incubation: If in a hay rick, it is
composed of hay; if against the trunk of a tree
that is covered with grey lichen, then the exterior
is formed of that plant; if in a bank, it is made
of green moss, but always invariably lined with

feathers: We have had them frequently on fir-trees and in banks; these were always of green moss. Those on trees are like a bunch of green moss, eight or ten inches in diameter, with a small hole on one side of it.

The wren does not, like other birds, begin at the bottom of the nest, and build upwards, but lays, as it were, the frame-work, and roofs it in, and then covers it all over, excepting a hole at the side for entrance. In this warm fabric the hen deposits from eight to sixteen eggs: they are a little larger than the golden-crested wren's, are white, and freckled with small spots of a reddish orange-colour. As soon as the young are hatched, the parents may be seen flitting to and fro every two or three minutes, with insects in their bills for their tiny brood, and this from morning to night during twelve or fourteen days together, with only a few hours' intermission in the course of the twenty-four. How astonishing that the little wing is not wearied by the exertion! How wonderful that sixteen bills in this dark recess should be recognised by the parent, and not one of its brood starved, and not one overfed! But the ways of Providence are wise as well as wonderful. Instinct, in some respects, seems less liable to error than the boasted reason of man. For we know

reason is often clouded and led astray by the pas-
sions, whereas instinct appears to act more immediately under the influence of that good and gracious Being who gave it.

This little warbler is reared, fed, and managed
as the redbreast; only, in a state of captivity,
there ought to be put, at one corner of the cage, a
small box lined and covered with cloth to keep
the bird warm in winter, with a hole in the box
towards the cage, to let the wren go in and out
at pleasure.

Description and Plumage.

Length between three and four inches; bill,
hair-brown; eyes umber-brown, or hazel; over
the eyes there is a streak of very pale wood-brown,
inclining to white; head, back, wings, and tail,
pale wood-brown, inclining to yellowish-brown :
all these parts are prettily barred with umber-
brown; throat, breast, and under parts pale wood-
brown,—the breast and sides slightly barred with
very pale hair-brown. Legs and feet pale yellowish-brown.

The male and female are very like each other;
therefore the best criterion to judge of a cock by,
is to hear him record or sing.

Thou little sportive, airy thing,
That trimm'st so oft thy yellow wing,
And, cheerful, pour'st thy varied lay
In sprighly notes, clear, rapid, gay :
As jocund in thy grated dome,
As thou at liberty did'st roam.

A captive born ! far from those isles
Where lavish nature richly smiles,
And where the broad Atlantic's wave
Thy native rocks and mountains lave ;
Less sweetly there, thy song is heard,
Than here, from liberty debarr'd.

E'en there, amid the palmy groves,
The gaudy finch in silence roves :
It seems as if thy carol taught
A lesson man has rarely caught,—
That cheerfulness can soothe our care
And teach us life's sad ills to bear.

Anonymous.

THE CANARY FINCH.

FRINGILLA CANARIA ; LINNÆUS.—LE SERIN DES
CANARIES ; BUFFON.

THE Canary, so universally admired for song
and plumage, and so well known in every part of

R. Scott Sculp.

THE CANARY BIRD.

Published by John Anderson Jun.r 55. North Bridge Street. Edinburgh. 1823.

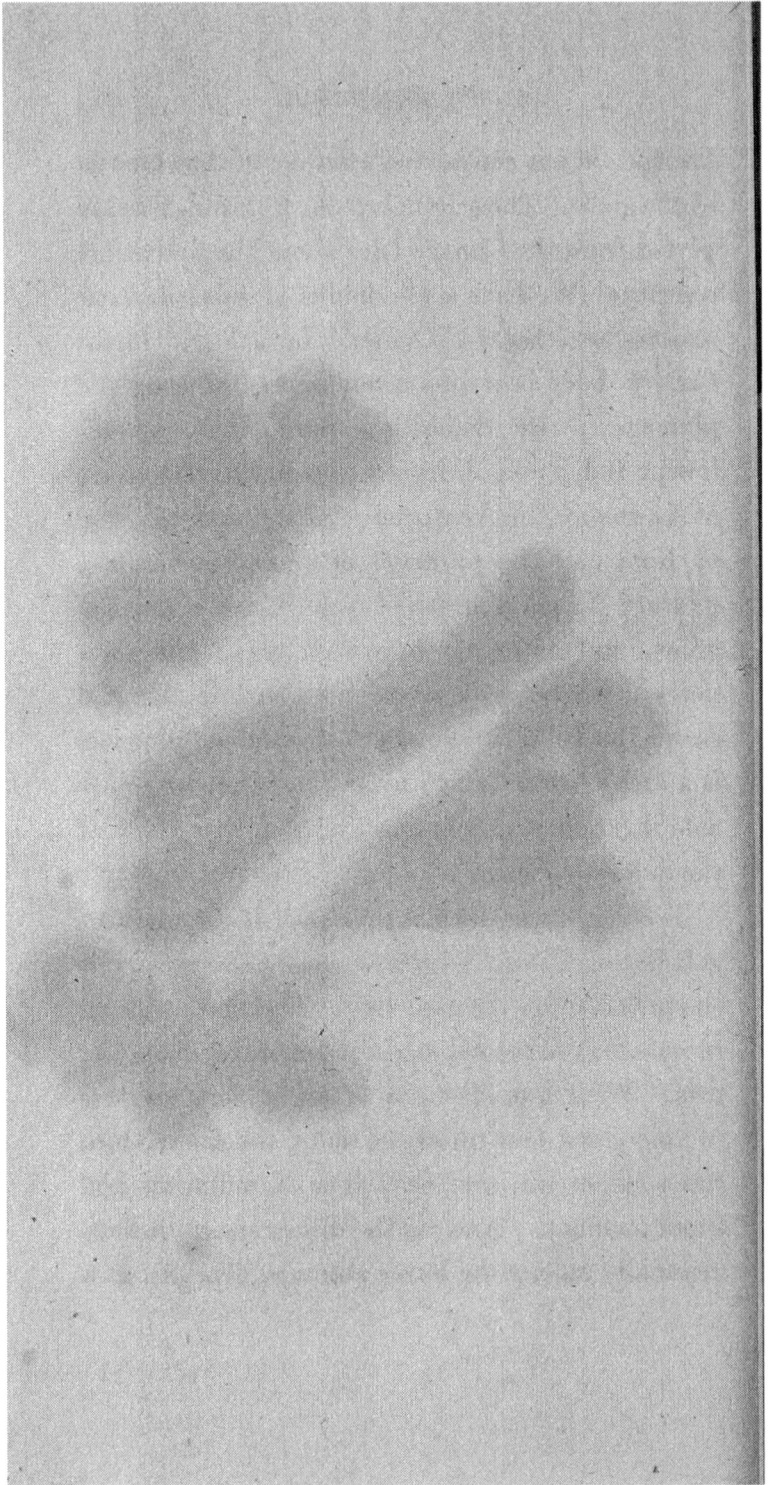

Europe, is not considered a native of that quarter of the globe. The original stock, it is said, was imported from the Canary Isles about the fourteenth century. We have some doubts of this; and our reasons are these :—The wild birds found in the Canary Isles bear less resemblance, in song and plumage, to the domestic canary, than the following indigenous European birds, *viz.* the siskin of Germany, the venturon of Italy, and the cini of France. The plumage of these species is a mixture of yellow, greenish-yellow, and yellowish-green, and very little brown or grey; and their notes are sweet, rich, and lively,—while the wild canary-bird has little or no song, and its plumage is a dingy grey. But we shall leave others to decide this query, and proceed with a description of the domestic canary.

The celebrated Buffon says :—" If the nightingale is the chauntress of the woods, the canary is the musician of the chamber. The first owes all to nature; the second derives something from our arts. With less strength of organ, less compass of voice, and less variety of note, the canary-bird has a better ear, greater facility of imitation, and more memory. And as the difference of genius, especially among the lower animals, depends, in a

great measure, on the difference that exists among
them, with regard to the perfection of their senses,
—the canary-bird, whose organ of hearing is more
attentive, more susceptible of receiving and re-
taining foreign impressions, becomes accordingly
more social, tame, and familiar. It is capable of
gratitude, and even of attachment: Its caresses
are endearing,—its little humours are innocent,—
and its anger neither hurts nor offends. Its education is easy: we rear it with pleasure, because we
are able to instruct it. It leaves the melody of its
natural note, to listen to the harmony of our voices
and instruments. It applauds, it accompanies us,
and repays the pleasure it receives, with interest:
The nightingale, more proud of its talent, seems
willing to preserve it in all its purity; at least it
appears very little to value ours, and it is with the
greatest difficulty it can be taught to repeat any of
our airs. The canary can speak and whistle: The
nightingale despises our words as well as our songs,
and never fails to return to its own wild wood-
notes. Its pipe is a master-piece of nature, which
human art can neither alter nor improve. That of
the canary is a model of more pliant materials,
which we can mould at pleasure. This last, therefore, contributes in a much greater degree to the

comforts of society. It sings at all seasons, it cheers us in the dullest weather, and even adds to our happiness: for it amuses the young, and delights the recluse; it charms the tediousness of the cloister, and exhilarates the soul of the innocent and the captive."

It is said there are upwards of thirty permanent varieties of this bird, which can easily be distinguished; and the number is increasing every year, and each kind has its admirers.

There are societies formed in London for promoting the breed of canary-birds, and a premium is awarded to the competitor that comes nearest what the society thinks perfection. A model is given out, the season prior to competition, which is represented by a coloured drawing accompanied with a technical description of the plumage, &c.— The limits of this treatise will not allow us to enter minutely into a detail of the different varieties of this bird, or the manner of breeding either them or mule-birds. For that purpose, we must refer the reader to several little works, published entirely on that subject, by amateurs and bird-fanciers in London; also to a treatise by Hervieux, and to the article *Serin* in the "*Nouveau Dictionnaire d'Histoire Naturelle.*" We shall, however, briefly mention a few observations relative to their manners, breeding, diseases, management, &c.

There are two distinct varieties of canary-birds: —the plain and the variegated, or, as they are technically called, gay or spangled, and jonks or jonquils. These two principal varieties are more highly esteemed by bird amateurs than any of the numerous mixtures which have sprung from them,— although birds of different feathers may have each their admirers; some preferring beauty of plumage, others excellence of song: but certainly the most desirable is, where both these qualities unite.— The plain, or jonquil birds, are of a rich, deep gamboge-yellow. The spangled or gay are variegated in their plumage,—the head, back, and wings of the last are of an agate colour, beautifully mottled. In some, the centre of the feather is dark—the margin pale; in others, the centre light—the margin dark; the primary quill-feathers and tail generally black, or tipped with black. The forepart of the head, the breast, and edge of the pinion on the cock, whether a plain or variegated bird, are always of a richer yellow than the same parts of the hen.

The dispositions of canary-birds are as various as their colours. Some are of a gay, sportive nature, delighting in sounds of mirth and revelry,

singing loudest and shrillest when they hear most
noise, and trying to drown, with their little pipe,
the voices and laughter of the company in the
room where their cages hang : These are most assiduous in assisting the hen to build her nest, and
even to hatch the eggs,—often sitting on them while
she is at liberty, and aiding her to feed the young.
They are clean, *debonnaire*, and easily tamed.

Others are sullen, intractable, and lazy of song.
Some cocks will destroy the eggs as soon as deposited; or, when hatched, will tear the young from
the nest and kill them in their rage. Others are
so indolent they will not build. The grey ones
never do ; and the person who superintends these,
must make a nest for them.

Some are so uncleanly as to draggle their feet
and tails; while others delight to bathe twice or
thrice a-day. We once possessed a jonquil cock,
that used to nibble at its prison-gate until it had
unlatched it; and then, escaping to the room,
would fly to the chimney-piece, and, placing itself
in any of the china ornaments, fluttered as if
in the act of washing itself, and continued
doing so till water was brought it; when, (like
the Beggar and the Barmecide,) it gladly re-
linquished the empty platter for the full one.

That same bird used to come, when called, to the
head or hand, run away with the ladies' thread
and needles, which it carefully deposited in one
corner of the cage; stopping, and often looking
round, as if it were for encouragement or applause:
But one of its favourite amusements was to perch,
and sing among the branches of some tall old myr-
tles which stood in the same window where hung
its frequently-open cage. Towards autumn, it
was usual to set its cage on a green terrace in the
garden, where at first, it (the canary,) hopped a-
bout with much wonder and seeming fear at so
large a demesne; at length acquiring courage to
try its little wing, it ventured first to the shelter
of a laurel, from whence it was speedily recalled,
and then a farther flight, to a willow grove close
by a stream, where the ephemeral insects that
rise from water seemed to afford it (as it darted
at them among the willow leaves,) a delicious ban-
quet. Towards evening, *Dickie* always returned
to his cage, the exit or entrance to which was at
last optional to himself. But we grieve to say
this license was finally his ruin; for the poor canary was found dead one Sunday, killed by a
young pointer, who, like *Dickie*, was a privileged
vagrant.

Various anecdotes might be given respecting the docility of these birds. We shall only notice the following, which took place at a public exhibition of birds. One canary, acting the part of a deserter, ran away, while two others pursued and caught him. A lighted match being given to one of these, he fired a small cannon, and the little deserter fell, and lay on his side as if dead: another bird then appeared, with a small wheel-barrow, for the purpose of carrying off the dead body; but, on the approach of the vehicle, the little deserter started instantly to his feet.

We have reason to think the canary-finch might be naturalized to our climate, as we remember to have seen, a few years ago, during the summer months, a pair of these birds flying about at liberty. Perhaps they were let loose by some person to try an experiment whether they would breed or not; and we supposed they had built a nest, from their being frequently observed flying in and out at the same spot, which was on a rocky precipitous bank at St. Bernard's Well, near Edinburgh. We do not know what became of these birds; for, we regret to say, they were cruelly disturbed by idle boys throwing stones at them.

Breeding, Management, &c.

Innumerable canary-birds are bred both as an article of commerce and for amusement in France, the Tyrol, Germany, and England. Those from Germany are usually spangled, and are esteemed the least, from their living only one or two years in this country, though the cock of this variety is an approved songster. Little is required in Britain for the rearing of these birds. A small breeding-cage often suffices; but, where a room can be allotted to the purpose, it ought to have shrubs in it for the birds to roost and build in, and plenty of water to drink and bathe in,—water being absolutely requisite for all birds. The light admitted into the room should be from the east, for the benefit of the morning sun, and the windows should have wire-cloth, that the birds may likewise enjoy the fresh air, so necessary for their health and preservation.

The floor of the apartment ought to be strewed with sand, or white gravel, and on that should be thrown groundsel, chickweed, or scalded rapeseed; but, when breeding, they must have no green food,—nothing, at that time, excepting hard chopped eggs, dry bread, or a cake kneaded without salt, and, once in two or three days, a few poppy-seeds.

When the eggs are laid, some bird-fanciers give the breeding birds plantain and lettuce-seed; but this must be done sparingly, and only for two days, lest this food should weaken them.

About the fourteenth or fifteenth day of April, they ought to be furnished with flax, soft hay, wool, hair, moss, and other dry materials, for making their nest, which usually occupies three days. The time of incubation is thirteen days; but, when the hen has sat eight or nine days, examine the eggs to see which are addle; hold them carefully, by the ends, against the sun or a lighted candle, and throw away the clear ones. Bird-fanciers, who wish the eggs to come out all on the same day, substitute an ivory egg, taking away the one that is deposited daily by the hen, until the last is laid, when the false eggs are removed, and the real ones put back into the nest. When canaries are to be reared by the stick, they must be taken from the mother on the eighth day, taking nest and all. Prior to this, the food for the young must be prepared, *viz.* a paste composed of boiled rapeseed, the yolk of an egg, and crums of the above-mentioned cake, mixed with a little water; and this must

H

be given every two hours; the paste ought not to be too wet, and, lest it should sour, it must be renewed daily until the nestlings can feed themselves. The hen canary has generally three broods in the year; but some will hatch five times in the season, each time laying six or seven eggs.

The weakest nestling is always the first to undergo the process of moulting, which takes place about five or six weeks after they are hatched; and this evil, which cannot be prevented, is often attended by serious consequences. When jonquil birds moult, it is frequently fatal to them; but the best palliative yet known is a small piece of iron put into the water that the birds drink: they must likewise be kept warm, cold being hurtful to all birds during this period. It is therefore less dangerous when birds moult early in the summer, as six weeks or two months generally elapse before they regain their strength and sprightly look.

All suffer from this evil, whether young or old; but, among all the varieties that are at present known, the white spangled are the strongest, and get easiest through when shedding their feathers. Hen canaries, after the sixth or seventh year, often die when moulting; and even the cock-birds,— though, from their superior strength, some may get through this malady, and continue occasionally to sing, and survive their mates four or five years; —yet, from this period, they appear dull and me- lancholy, and gradually droop, till they at last fall a victim to this evil.

If it is proposed to rear gay birds, the cock and hen should be of the same clear deep yellow ; and, if mottled birds are required, both parents must be mottled : When a gay and a fancy bird are matched, the offspring are termed mule-birds, because they are irregularly mottled in their fea- thers, and therefore of no value for plumage, though they may be equally good for singing.

A spangled, or fancy bird, ought to have the crown of the head of a pure yellow, or pure white, divided down the middle. A single feather of grey on the crown deteriorates from perfection, while on the back, wings, and tail, there must be no yellow : The more he is mottled on the upper parts of his plumage, exclusive of the crown, the more highly will the fancy bird be prized. These marks hold good in the hen as well as in the cock. There are two accidental varieties in fancy birds, known by the name of meally, and junks :—The former (or meally) those whose crown and under parts are of a white or *pale* yellow; the junks, whose crown and lower parts are of a *deep* yellow.

Some bird-fanciers pair the spangled or French

canary cock-bird with a meally hen, as the young are often highly and curiously spangled. From such a union many beautiful varieties have sprung.

Diseases.

The most common cause of disease in birds is a superabundance of food: When they breed in a cage, they often eat to excess, or take the succulent food intended for the nestlings.—This brings on either repletion or inflammation. Overfeeding is likewise often fatal to the young, and goes by the technical name of a surfeit: in this case the intestines descend to the extremity of the body, and are seen through the skin, while the feathers on the part affected fall off, and the poor bird, after sitting a few days by its untasted food, pines and dies.—Medicine is vain if the disease is far gone; but, by putting the sick birds in a separate cage, and letting them have nothing but water and lettuce-seed, this cooling diet sometimes saves a few out of many. The greatest care, therefore, is necessary to prevent too high feeding. The best food for birds brought up from the nest, by the stick, is boiled rapeseed and a little groundsel.

Canary birds are also subject to epilepsy, to asthma, to ulcers in the throat, and to extinction of voice. The cure for epilepsy is doubtful.—It is alleged that, if a drop of blood falls from the bill, the bird shortly recovers life and sense; but if touched, prior to the blood falling of itself, it occasions death. One thing, however, is certain, that, if they recover from the first attack, they frequently live after it many years, and sing as well as if they had never experienced a fit; and probably all might survive were a slight wound given them in the foot. The asthma is cured by plantain, and by hard biscuit soaked in white wine. Ulcers, like the surfeit, proceed from too much or too succulent food, which brings on inflammation in the palate; it must be cured by cooling food, such as lettuce-seed, with water, in which bruised melon-seeds have been steeped.

For extinction of voice, their food ought to be, hard yolks of egg, chopped down with the crums of bread; and, for their drink, put a slice of liquorice-root, or a blade of saffron, in the water. In addition to these evils, canary birds are often infested with a small insect owing to their being kept dirty. To avoid this, they ought to have plenty of water to bathe in,—a new wooden cage, and, if covered, it ought to be with new cloth, the old being liable to attract moths,—and the food and seed should be sifted and washed. If these attentions are troublesome, they are nevertheless necessary if it is wished to have a thriving bird. When wild, it has already been remarked that all birds require water; and, for the canary, bathing is so requisite, that, if a saucer or cup of snow be put in their cage, they will flutter against it with apparent delight, even during the most severe winters.

Description and Plumage.

The length of this elegant little bird is about five and a half inches; the bill very pale flesh-red, passing into reddish-white; eyes chestnut-brown; the whole plumage is of a rich, deep primrose yellow colour, inclining to gamboge-yellow; edges of the quills sometimes yellowish-white; legs and feet the same colour as the bill. This is the description of a canary bird by bird-fanciers called a "jonquil," or "gay bird;" the male of which is distinguished from the female by his plumage being deeper in the yellow round the bill and eye, and on the breast and edge of the wing, particularly the false wing; he is also rather less, and more slender in form towards the tail.

The fancy, or spangled bird, has a tuft of feathers on its head, with a groove in the centre,—the feathers of the tuft bending or inclining from the middle to the sides.—This tuft is of fine gamboge-yellow; the throat, breast, and under parts the same; and the neck, back, and wings are beautifully waved and mottled with different tints of blackish grey, passing into very pale purplish-grey, or what bird-fanciers term agate colour; the tail is blackish-grey. The bill, legs, and feet are of the same colour as those of the gay canary. The cock-bird is altogether much brighter in his plumage, but best distinguished from the female by his shape, which is more slender towards the tail.

Song.

It is difficult, at the present day, to distinguish what originally were the natural notes of this bird, as most canaries are now brought up under the nightingale, the woodlark, or some other excellent song-bird. Some canaries begin with the notes of the nightingale, and end with those of the woodlark: Others commence with those of the woodlark, and conclude with the full cadence of the nightingale.—But so varied are the songs of these birds, that it is impossible to give a correct description of any of them. We shall merely notice, that, although these birds possess many of the notes of the nightingale and woodlark in their song, yet their warble wants that rich, mellow, plaintive tone, that the songs of the two last mentioned birds possess: The difference may be compared to the same air played on the German flute and on the common fife; for, though the song of the canary is rich, varied, cheerful, expressive, and continued—it is shrill, and often too loud and piercing for very fine-toned melody.

THE SISKIN;

OR ABERDEVINE.

FRINGILLA SPINUS; LINNÆUS.—LE TARIN; BUFFON.

SISKINS, though not equal to canaries as song-birds, yet bring as high prices, because bird-fanciers are always anxious to possess them, for the purpose of pairing them with canaries. They are healthy, mild, and docile birds—and, when paired with ca-naries, their progeny generally inherit the same good qualities, and therefore they are highly prized by amateurs.—Besides, no other bird pairs so readily with the canary: Whether it be the hen siskin with the cock canary, or the cock siskin with the hen canary—the male and female of each species pair equally well. This is not the case with the goldfinch, chaffinch, or linnet: it is only the males of these birds that pair with the hen canary:

H 2

and this leads us to think, that, if the siskin is not the wild canary, or stock-bird, it is a species that approaches so close to it as almost to appear only a marked or distinct variety. These birds are common in Europe, though rather rare with us, and said to be only winter visitors.

Montagu mentions, that, "in the month of December 1805, a small flock of these birds were seen, busy in extracting the seed from the alder trees in the south of Devon; several of which were shot. The weather was severe, and a heavy fall of snow succeeded." And Bewick remarks, that one, which he kept many years in a cage, had a pleasing and sweetly-varied song, and that it imitated the notes of other birds: It was caught on the banks of the Tyne. About London, the siskin is called the aberdevine by bird-catchers, who occasionally take a few of these birds. In all places they are migratory, but irregular in their migrations. In Germany they appear about October, when they do a great deal of damage to the hop plantations; and the places where they have been are easily known by the number of leaves that are found lying on the ground. They visit France during the vintage, and even earlier in the year, when they injure the blossoms of the

apple-trees. Buffon says, that " immense flocks
of these birds appear every three or four years."
They are said to fly very high, and may be heard
before they are seen. It is curious that their
nests are so rarely found; nor is it certain where
they breed, but it is supposed to be in mountainous forests.

On the banks of the Danube, Kramer remarks
that thousands of young siskins are seen, which
have not yet dropped their nestling feathers. These
birds surely must have been bred there, or at
least not far distant. Sepp has delineated the nest
as "placed in the fork of a tree, built with dry
bent mixed with leaves, and amply lined with
feathers; the eggs, three in number, are of a dull
white." These birds are of so mild, gentle, and
docile a disposition, that they become quite tame almost immediately after they are taken. They may
be taught many pretty tricks, such as to open the
door of their cage, draw up their food and water,
and come to the hand to be fed at the sound of a
little bell or a whistle. Their food is the same as
that of canaries, and they are managed in the
same manner.

Description and Plumage.

Length rather more than five inches; bill reddish-white, tipped with blackish-brown; eyes, umber-brown; head, greenish-black; over each eye a pale streak of dingy primrose-yellow; neck, back, wings, and tail, oil-green,—paler, and more on the yellow, on the lower parts of the back towards the tail-coverts. The feathers of the back and wings are streaked down the middle with a tint formed of blackish-green and hair-brown; sides of the head, throat, breast, and under parts, pale wax-yellow, inclining to sulphur-yellow; middle of the parts below the breast, very pale wine-yellow, passing into white; across each wing are two bands of primrose-yellow, and between them one of black; part of the quills and tail edged with pale gamboge-yellow; legs and feet pale flesh-red. The head of the female is of a brownish-colour, inclining to grey where the male is black; and all the rest of her plumage is of a more dingy colour. The siskin nearly resembles the canary, termed the green variety, only it is a little less, the tail being rather shorter in proportion.

R. Scott Sculp.

GOLDFINCH.

Published by John Anderson Junr. 55. North Bridge Street. Edinburgh. 1823.

" Hid among the op'ning flowers
Of the sweetest vernal bowers,
Passing there the anxious hours
In her little mossy dome,

Sits thy mate, whilst thou art singing,
Or across the lawn seen winging,
Or upon a thistle swinging,
Gleaning for thy happy home."

Anonymous.

THE GOLDFINCH,

THISTLE-FINCH, OR GOUDSPINK.

FRINGILLA CARDUALIS; LINNÆUS.—LE CHARDON-
NERET; BUFFON.

THIS beautiful and delightful warbler, in its
disposition, possesses many of those qualities that
even in man would be thought agreeable. It is
amiable in its affections, mild, docile, cheerful,

and contented; and its actions are sprightly, sportive, and graceful. The plumage of the goldfinch
is rich and varied, and, what is singular, its song
is sweet and cheerful;—we say singular, for, where
Nature has given richness and brilliancy of plumage, she seems almost invariably to have denied
sweetness of song. The goldfinch is easily tamed
and easily taught, and its capability of learning
the notes of other birds is well known; but the
tricks it may be taught to perform are truly astonishing. A few years ago, the Sieur Roman exhibited his birds, which were goldfinches, linnets,
and canaries: "One appeared dead, and was held
up by the tail or claw without exhibiting any signs
of life; a second stood on its head with its claws
in the air; a third imitated a Dutch milk-maid
going to market, with pails on its shoulders; a
fourth mimicked a Venetian girl looking out at a
window; a fifth appeared as a soldier, and mount-
ed guard as a sentinel; and the sixth acted as a
cannoneer, with a cap on its head, a firelock on its
shoulder, and a match in its claw, and discharged
a small cannon. The same bird also acted as
if it had been wounded.—It was wheeled in a bar-
row, to convey it as it were to the hospital; after
which, it flew away before the company: the

seventh turned a kind of windmill; and the last bird stood in the midst of some fire-works which were discharged all round it, and this without exhibiting the least symptom of fear."

They may also be taught to draw up little buckets or cups with food and water. To teach them this, there must be put round them a narrow soft leather belt, in which there must be four holes,—two for the wings, and two for the feet. The belt is joined a little below the breast, where there is a ring to which the chain is attached that supports the little bucket or cup: We have seen both the goldfinch and lesser redpole perform this action, but in a different manner. Their cage had no wires,—only a back-board, a bottom-board, and one perch : to one foot of the bird was attached a light slender chain, which allowed it more exercise than it could have had in the common wire cage; at the outer edge of the bottom board was a ring through which ran the chain, to each end of which were fastened the little buckets that held the food and water, which the bird drew up with its foot and bill; and, as one bucket was drawn up, the other sunk,—thus lessening the difficulty, and lightening the task.

It appears to be a vain bird; for, if a looking-glass is placed before it, the reflection of its own gay feathers seems greatly to delight it.

The goldfinch is a long-lived bird. Willoughby mentions one that lived twenty-three years in a state of confinement: But they sometimes turn white with age, particularly the red and yellow parts of their plumage.

In a wild state these birds assemble towards autumn, and pillage gardens, &c. They fly in large flocks. The flight of the goldfinch is low and equal, not in jerks or bounds. Their natural food is seeds of different kinds, such as thistle, teasle, plantain, chickweed; also cherries, guignes, (or mazzards,) pears, and apples; and they always have the discrimination to choose the best fruit. During winter they may be seen near high-roads, brushing with their wings the snow off plants to get at the seeds. It is doubted by some whether they ever feed on insects, even the young; but we are rather inclined to think that the young are fed with caterpillars, grubs, &c. which the parent birds find amongst the blossoms of trees where they build their nests.

Of the Nest and Eggs.

In spring they haunt gardens and orchards, or
plantations and shrubberies near gardens, where
they breed. The nest is artfully concealed among
the leaves and blossoms of apple and pear-trees,
generally placed on some limber branch, where
cats, &c. cannot get at it :—In shrubberies, they
select the tops of the thickest evergreens, or high
hedges, to build their nests in. The nest is uncommonly neat, perhaps the prettiest structure of
any European small bird : It is somewhat like the
nest formed by the chaffinch, but still more neat,
more compact, and rather less ; it is composed of
bents and moss, mixed with wool and lichen or
that whitish green moss so often seen covering the
trunks and branches of old trees, and lined with
hair, covered with thistle-down and cotton from the
catkins of the willow, the cannach, or cotton grass,
&c. ; the eggs, five or six in number, are of a
bluish-white, sprinkled with a few small spots of
a reddish orange-colour. The hen is a very close
sitter.—Storms of wind, rain, or even hail, will
not drive her off the eggs to seek shelter; and the
cock-bird is tender and attentive to her while thus

employed,—never wandering far from the spot where all his care and joy are centered: he brings her food, and soothes and cheers her with his song.

The young are rather delicate. They ought to be pretty well fledged before they are taken; but it is unnecessary trouble to rear them from the nest, as the old are so easily tamed; unless we wish to pair them with canaries, in which case the goldfinch must be reared from the nest.

Of the Young and their Food.

The young of the goldfinch, like all other song-birds, must be kept very clean and dry. Damp is injurious to all land birds. Their food ought to be loaf-bread boiled in milk diluted with a little water, and made into a paste and mixed with a little flour of canary-seed. They ought to be fed very early in the morning. Give each bird three or four small bits on the end of a stick, and that every two hours till sunset. In about a month they will be able to feed themselves. They are then, as well as adult goldfinches, to be fed and managed as canary-birds. The young, for some time after they leave the nest, are brownish about the head, and then by some bird-fanciers are called grey-pates.

Diseases.

Goldfinches are rather hardy and healthy birds; but, should they droop, put a little saffron in their drink, strew their cage with clean gravel, and put some pounded chalk amongst it; also, give them some chickweed, groundsel, and thistle-seeds; if their claws get too long, they must be cut, and, if troubled with epilepsy or the pip, the birds must be treated in the same manner as canaries. It is said too much hempseed is the cause of epilepsy; we know, also, it injures the beauty of the plumage.

Song.

Their native notes are sweet, cheerful, and sprightly, at times chattering or twittering, which makes an agreeable variety. If young goldfinches are brought up near a canary bird, wood, or titlark, they readily acquire their notes; but it is said the goldfinch learns the song of the common wren sooner than any other.

Description and Plumage.

The length of this elegant species is about five and a half inches; the bill straight—reddish-white, at the point inclining to brownish-black. Eyes clear umber-brown; forehead and chin brilliant arterial blood-red; top of the head bluish-black,—the black extending from the head round the cheeks, and nearly to the throat; cheeks and hind parts of the neck yellowish-white; back pale yellowish-brown; breast pale wood-brown; under parts inclining to dingy yellowish-white; quills dull greyish-black, crossed with a large bar of bright gamboge-yellow; tail greyish-black; it and the quills tipped with dull yellowish-white; legs and feet very pale flesh-red, inclining to white.

R. Scott Sculp.

GREY LINNET.

Published by John Anderson Junr 55. North Bridge Street, Edinburgh, 1823.

The Lintie, on the heathery brae,
(Whare lies the nest amang the ferns,)
Begins its lilt at break o' day,
And at the gloaming hails the sterns.

I wadna gie the lintie's sang,
Sae merry on the broomy lea,
For a' the notes that ever rang
Frae a' the harps o' minstrelsy !

Mair dear to me whare buss or breer
Amang the pathless heather grows,
The Lintie's wild, sweet note to hear,
As on the ev'ning breeze it flows.

Anonymous.

THE LINNET.

GREY LINTIE.—BROWN LINNET.

FRINGILLA LINARIA; LINNÆUS.—LA LINOTTE;
BUFFON.

THIS charming little bird, though plain in its
plumage, is in its shape elegant; and its song is
sweetly varied;—now rich and sprightly, now ten-

der and expressive. Even in a cage, its little pipe
conveys to the mind the pleasant feeling that
the tiny captive is cheerful, contented, and happy.
It is therefore highly prized as a song-bird, and
there are some who think its notes equal to those
of our finest warblers.

These birds are not only cheerful and docile in
their disposition, but social, and fond of society.
In summer they breed in numbers near each other
on the wilds and heaths; and in winter they assemble in flocks of hundreds together, and continue so till they begin to pair in spring. In
autumn they frequent stubble-fields, and, if snow
is on the ground, they repair to farm-yards in
quest of food; in their flights, they rise, wheel,
and alight, (as if by mutual consent,) on the same
field, hedge, or trees. They build their nests in
wild heathy places, or furzy brakes, amongst fields,
often near farm-houses and villages. This the
redpole never does, being a wild and shy bird. The
nest of the linnet is composed of bents mixed with
moss and wool, and neatly lined with white or
grey horse hair. The eggs, four or five in number, are of a reddish-white colour, speckled with
small spots of reddish-orange. The young may
be taken after seven or eight days: they are easily

reared; and ought to be fed with scalded rape-
seed and loaf-bread, boiled with milk and water.
The bread is then taken out and mixed with the
rape-seed, and made into a paste : with this they
must be fed every two hours till they are able to
feed themselves, which generally takes place in a
month or five weeks: they are then fed with rape
and canary-seed.   In Scotland, boys rear thousands of these birds on oatmeal and cold water,
(called drammock,) and with this alone. The old
birds, when taken, are easily tamed, and their food
is the same, *viz.* rape and canary-seeds.   When
these birds are affected with disease, they are
treated in the same manner as goldfinches, canaries, &c.

Description and Plumage.

The grey linnet is a very elegant-shaped bird.
Length about five inches and a half; bill bluish-
grey; eyes umber-brown; head, neck, back, wings,
and tail, pale umber-brown, with deeper shades of
the same colour down the centre of each feather;
spurious wings brownish-black, edges of the greater
quills brilliant snow-white, forming a large space
of white in the wing; outer edges of the tail feathers also bright snow-white; legs and feet yellow-

ish-brown. The white on the wings and tail at
once distinguishes the cock from the hen,—the
white on her wing and tail being more dingy and
less broad; by this mark also the cock linnet may
be easily known from the males of the redpole
and twite,—their wings being even more dingy
than those of the hen linnet's.

Song.

The warble of the linnet is sweet, sprightly, and
varied, and conveys to the mind a feeling of cheerful gaiety. It has none of the twittering, chattering quick notes of the goldfinch, but is altogether more plaintive; yet the tones are not
sad, like the mellow, plaintive notes of the nightingale, the redbreast, and the blackcap; the cadence,
from beginning to end, is a sweet and varied melodious strain. In a cage, in fine weather, they,
like the canary, are constantly singing, and they
will continue to sing for nine months in the year.
The arrangement of the notes of their warble is
different, and their song is altogether perhaps
sweeter than those of the redpole, redbreasted
linnet, or twite; and the notes of these three last
mentioned birds differ from each other, but more
in the arrangement of the notes than in the superiority of their songs.

THE GREATER REDPOLE.

FRINGILLA CANABINA; LINNÆUS.—LA GRANDE
LINOTTE DES VIGNES; BUFFON.

BEFORE proceeding with the description, &c. of
the Greater Redpole, we may here remark, that
we intend giving only short notices of it, and the
following birds, viz.—The Red-breasted Linnet,
Twite, and Lesser Redpole, as these excellent song-
birds are so nearly allied to the preceding species,
or grey linnet, in habit and manners, that an
extended description might appear superfluous.
Some authors think the greater redpole and red-
breasted linnets are merely grey linnets in a different plumage. But we are rather inclined to
consider them all as distinct species.——*See Introduction.*

The greater redpole is somewhat longer than
the grey linnet, being rather more than five inches
and a half. It has a wild, sweet, and varied war-

I

ble.   It builds in wild and lonely
furze and heath, or in low bushes
Its nest is formed of bent and
with fine hair, cotton-grass, or w

The eggs, four or five in num
between bluish and greenish-whi
purplish-brown spots, that form
large end.

The redpole never builds near
men, like the linnet. It is found, i
mer, among the wild and heathy
the Scottish and English borders,
of Yorkshire.   There, in a state
little warbler " wastes the swee
upon the desert," amid the fui
o'er its nest.

The young for a considerable
female, and, if caught in that sta
the red feathers which distingui
wild.   In winter it migrates sor
The young are reared, and they

rather longer than the grey linnet, being
above five and a half inches,—bill thick at
—the upper mandible pale hair-brown,
the point,—the under one paler, inclining
—eyes umber-brown, or hazel,—over ea
streak of yellowish-grey, passing into w
same under each eye,—on the forehead
spot of rich deep arterial blood-red,—b
same, but streaked downwards; differing i
spect from the red-breasted linnet, which
fully waved horizontally,—the hind-head
ash-grey, streaked with umber-brown,
dingy ash-grey,—chin dingy yellowish-
upper part of the breast, above the red,
lowish-white, spotted with umber-brown,-
a very pale chestnut-brown, passing into d
lowish-white towards the middle of the low
—the back and wings chestnut-brown, deep
middle of most of the feathers,—spurio
black,—quills edged with dingy-white, pas
wood-brown—tail brownish-black, and s

RED-BREASTED LINNET.

ROSE LINTIE.

Description and Plumage.

This bird is rather longer then the redpole. The bill is bluish-grey, inclining to pale wood-brown, blackish at the point,—eyes umber-brown,—back and wings chestnut and yellowish-brown; some of the feathers slightly marked in the middle with umber-brown,—quills bluish-white,—tail brownish-black,—outer feathers edged with bluish-white—breast pale lake, or carmine-red, beautifully waved horizontally.

Below the breast, and downwards, very pale wood-brown, inclining to ash-grey, and passing into yellowish and bluish-white,—legs and feet dingy yellowish-brown. The plumage of the back and wings of the redpole are not so much mottled as those of the grey-linnet, and this bird's are less so than the redpole's, and more of a brown cast than either. The treatment of these birds, young and old, is the same as that of the goldfinch or the linnet.

The red-breasted linnet is more rare than the grey-linnet; but much more common than the greater redpole. It builds in the same situations as those birds, but oftener in furze and bushes in ravines, and amongst craggy places; from which circumstance some call it the rock-linnet.

The nest is composed of withered grass, moss, and wool; lined with hair and cotton-grass, or thistle-down. The eggs, from four to five in number, are of a bluish-white, sprinkled with reddish-brown spots.

The song of the red-breasted linnet is very rich and varied; and it pours out its note, o'er the solitary wilds, in a strain of sweet and cheerful melody.

TWITE,

OR MOUNTAIN LINNET.

FRINGILLA MONTIUM; GMEL.—LINARIA MONTANA;
BRIS.—LA LINOTTE DES MONTAGNES; BUFFON.

THESE birds are the wildest of the genus. They
are more shy than even the greater redpole. The
most solitary fells and moors, and the bases of
mountains remote from the habitations of man,
are their usual haunts; and there they build their
nests: But, like the redpoles, migrate southward in
winter. In spring and summer they are found a-
mong the hills, and wild heaths, and moors, along
the Scottish and English borders. The nest is
formed of heath and dry grass, and lined with
down and feathers: And the eggs are of a pale
bluish-green colour, inclining to white, and sprinkled with spots of chestnut-brown. This bird is
managed in every respect as the grey-linnet.

Description and Plumage.

Length five inches,—bill short, thick, and of a
pale flesh-red, inclining to reddish-white,—nostrils

covered with short hairy feathers,—eyes umber-brown; there is a small space above and below the eyes, of a tint between pale wood-brown and pale reddish-orange,—throat and fore part of the neck the same tawny tint, faintly marked with spots of a deeper shade of the same colour,—breast and sides pale dingy wood-brown,—lower parts pale greyish-white,—hind parts of the neck and sides dingy yellowish-white, spotted with umber-brown, —the back, wings, and tail, umber-brown; deepest on the back and tail,—feathers of the back edged with pale wood-brown,—greater wing-coverts tipped with dull white; primary quills slightly edged with pale wood-brown; secondary quills edged with yellowish-white,—tail, somewhat forked,—outer edges of the outer feathers of it margined slightly with white, passing into dingy white towards the end,—the lower part of the back or rump is of a beautiful pale lake-red, inclining to peach-blossom red,—feathers behind the legs, yellowish-white, with a streak of brownish-black down the middle of the centre feather,—legs and feet hair-brown.

The female wants the red on the lower part of the back.

LESSER REDPOLE.

FRINGILLA LINARIA; LINNÆUS.—LE SIZERIN, OU
LA PETITE LINOTTE DES VIGNES; BUFFON.

THIS is a pretty little bird, nearly allied to the preceding species. Perhaps its natural song is rather inferior to theirs; yet, with care, it may become a good cage bird: For that purpose it ought to be placed beside a canary or linnet.

In spring and summer they frequent the northern counties of England and southern borders of Scotland, where they breed. The nest is composed of bent, moss, and willow-down; and lined with the down of the willow, thistle, or cotton-grass. The eggs are of a bluish-white, tinged with very pale bluish-green, and freckled with reddish orange-coloured spots, principally at the large end. Mr Pennant found the nest of this bird on an alder stump: It was lined with hair, but, in other respects, resembled that mentioned above; and the eggs, four in number, were the same as those already described. He also says: " The bird was so tenacious of her nest as to suffer us to take her off with our hand; and we found, after we had released her, she would not forsake it." In winter these birds migrate southward, and are sometimes taken about London by bird-catchers; where they are known by the name of stone redpoles. The lesser and greater redpoles are subject to considerable variation in their plumage; but, in all their changes, they never lose the appearance of a distinct species from the grey-linnet, the red-breasted linnet, and twite. The lesser redpole is reared and managed in the same manner as the linnet and goldfinch.

Description and Plumage.

Length about five inches,—the bill of a pale yellowish-grey, inclining to yellowish-white,—eyes umber-brown,—on the forehead there is a spot of red, of a tint between arterial blood-red and auricula-purple,—hind-head, neck, back, wings, and tail, pale umber-brown, the feathers edged with very faint chestnut-brown, inclining to wood-brown;—chin brownish-black,—throat and breast tinged with a hue between pale lake-red and crimson-red,—sides streaked with hair-brown, lower parts yellowish-white,—quills hair-brown, edged with wood-brown,—tail feathers margined with pale wood-brown,—legs and feet hair-brown. In some, the lower parts of the back is tinged with pale crimson.

Where the chaffinch rests its wing,
  'Mid the budding trees so gay,
Still, anon, it loves to sing,
  Merrily, its roundelay.

Lo! on yonder branchlet hoar,
  Twin'd with honeysuckle round,
Curiously bestudded o'er,
  Lurks a nest by ivy crowned.

In that little mossy nest,
  Hid from truant school-boy's eye,
Warm, beneath the shilfa's breast,
  Twice two speckled spheroids lie.

See! her mate, as nigh the tree,
  Chaunting oft at break of day,
Still proclaiming, merrily,
  Merrily, his roundelay!

Anonymous.

THE CHAFFINCH.

SHILFA.——TWINK.

FRINGILLA CŒLEBS; LINNÆUS.——LE PINÇON;

BUFFON.

THIS airy and elegant little warbler is the un-
molested favourite of every cottage child; and,

THE CHAFFINCH.

Published by John Anderson Jun.ʳ 55.North Bridge Street,Edinburgh.1823.

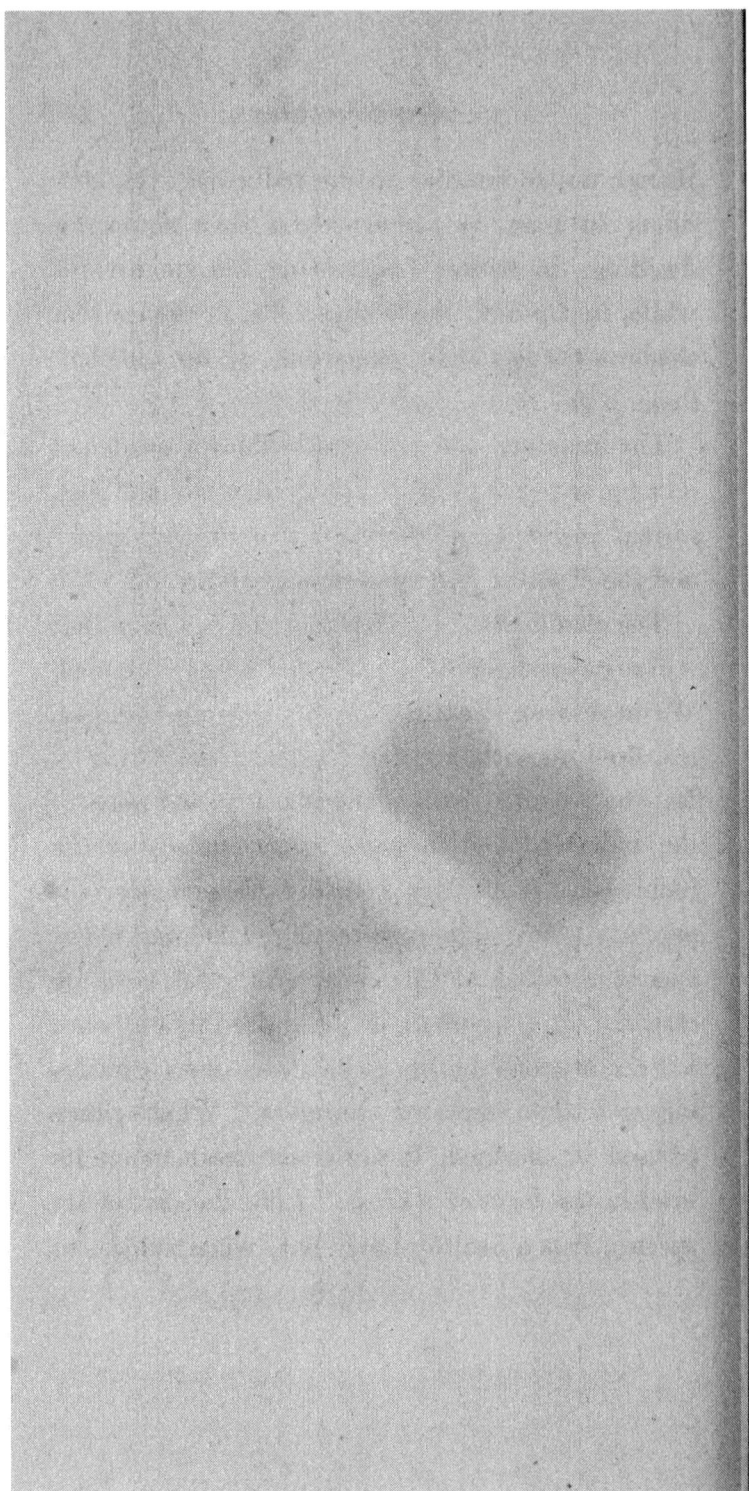

though not so familiar as the redbreast, yet, con-
fiding in man, it rather courts than shuns his
dwelling, in winter frequenting the stack-yard;
while, in summer, the honeysuckle, or the ivy that
shadows the bower in his garden, is, for the chaffinch, a chosen residence.

The imagined and universally known burden of
its song is repeated by every country boy and girl,
so that in Scotland " *the wee, wee, drucken sowie*"
and the " shilfa" are synonymous terms.

The chaffinch is not often confined in a cage, but
is of so gay and gentle a nature that it is easily tamed.
We have seen one that was taken from the nest,
and bred up with a redbreast and a mule-bird.—
Its song partakes both of the richly-varied notes of
the mule-bird and the wild sweet melody of the
redbreast. It displays considerable sagacity, apparently knowing its own name, by coming when-
ever it is called, and perching on the finger of its
master. It delights in being noticed by any one;
but the attentions of its master seem more gratify-
ing to it than those of strangers. When either
pleased or alarmed, it raises the feathers on its
head in the form of a crest. Like the rest of its
species, it is a healthy bird; but, when subject to

diseases, they are the same as tho
finch, and must be treated in a sim

 ˙ The chaffinch with us is a static
in Holland it is known to migr:
numbers of male chaffinches migrat(
to this island in winter.

Of the Nest and Eggs

The nest of the chaffinch is]
artfully contrived. Except the nes
finch and golden-crested wren, no
beauty; and the very spots wher(
build are generally romantic—in
branchlet, or among the decaying
old apple trees,—in the topmost]
hedges; and once we found one or
a tree overhanging a public road, a
footpath. The nest is composed (
withered grass, and slender sticks, (
ven with the grey lichen, (common

to be distinguished from the moss-grown
on which it is placed, and therefore very in
tible to a common eye.   It is lined with s
hair, and feathers, and contains from fou
eggs of a dirty reddish-white, streaked an
led towards the larger end with an umbe
In general only four birds are hatched, a
are fed, when young, with caterpillars, smal
aphidæ, &c.

Description and Plumage.

The plumage of the chaffinch is beaut
not gaudy.   Length rather more than fiv
half inches; bill, pale bluish-grey tipt wit
eyes, umber-brown; forehead, very deep
black; crown of the head, back part, and
the neck, ash-grey, inclining to pearl-grey
the eyes, cheeks, sides of the neck and thi
reddish-orange; breast, very pale aurora-r
ing into dingy peach-blossom red; low

white on the coverts next the scapularies; secondary coverts black, tipt with yellowish white; so
that, when the wings are closed, two bars of white
are seen.

··.The lower white bar, in some places, is tinged
with gamboge-yellow; primary quills, greyish
black, edged with bright sulphur-yellow, slightly
tipt with dingy yellowish-white at their points; tail
greyish black, outer feathers of it obliquely margined with snow-white, the white extending broad
at the base; outer web of the exterior feather wholly
white; legs and feet a tint between yellowish-
brown, and pale hair-brown. The female, on the
upper parts, is of a dingy hair-brown, inclining to
pale oil-green; breast and lower parts, dull yellowish white, inclining to wine yellow, and passing into wood-brown; her wings, like those of the
male, are also barred with white, but less broad
and less brilliant.

, The chaffinch is a lively bird, tripping lightly
along the ground on half-spread wings, continually
in motion, rarely perching or sitting, except when
at rest for the night.

Song.

The natural note of the chaffinch is so unvaried, and so frequently repeated, that, although very sweet in the open air, when heard at intervals among the richer melody of others of the feathered choir, it possesses little interest in a domestic state; but, if bred from the nest, the chaffinch easily acquires the note of any good warbler that they are brought up with; such, for instance, as the bird we have noticed in the foregoing page, whose song is both melodious and varied, being composed of part of the redbreast's and part of the mule-bird's song. In a wild state, the chaffinch begins its gay, brief carol early in April, and continues to sing until midsummer, at which time " *the wee, wee, drucken sowie*" is entirely laid aside for a mere chirp.

During rain, they make a curious kind of whirring noise. The note of the female, like that of the male, after the summer solstice, has a sort of metallic sound, which Montagu expresses by the repetition of the word " twink," from whence comes one of the English names of this bird.

Beside a little straw-built shed,.
An ancient garden sloping lay,
To meet the sun from Ocean's bed,
And catch his last departing ray!
Within that garden's sunny bound,
The sweet birds chaunted morn and eve;
And there the cottage children found
A fabric such as elfins weave.
It was a curious mossy cell,
Woven with twigs, and grass, and hair,
And, 'mid the moss six nestlings dwell
Concealed by apple-blossoms fair.
" 'Tis Bully's nest !" Bethia said,
" His head of glossy jet I spy,
His downy breast of softest red;
Poor bird! I hear his whooping cry."

Anonymous.

THE BULLFINCH.

RED-HOOP, ALP, OR NOPE.

LOXIA PYRRHULA; LINNÆUS.—LE BOUVREUIL;
BUFFON.

THIS species of grossbeak is a very beautiful
and docile bird. Its natural song is not powerful,

R. Scott Sculp.

BULLFINCH.

but very sweet: early in the morning, or when
undisturbed, its warble is low, soft, and melodious. Its aptness to learn what it hears is astonishing; for it may be taught to pipe tunes, to
articulate words, and even to whistle in parts.
Two of these birds have been known to sing or
whistle a duet, keeping accurate time and tune.

" Piping Bullfinches" is a common phrase for
the taught birds exhibited in this country. The
Germans are the best teachers of these birds, as
well as of many other species. One which a lady
(north of the Tay,) bought from a German bird-
dealer, because of the excellence of its song, with
which she had been much delighted, was no sooner
in her possession, than it became entirely mute,
and, though apparently in perfect health, neither
voice nor instrument could induce it to sing. A
Hanoverian officer, who happened to be in that
part of Scotland where the lady resides, saw the
bird, and whistled to it various waltzes, but in
vain,—Bully was still silent; at last the Hanoverian
bethought him of an air he had heard a bird-
catcher sing in Germany, and, whistling the first
bar, the bullfinch instantly finished it.

Another that we know of now, (1823,) in Edinburgh, not only sings delightfully, but performs

several curious tricks. When its mistress is at work, it flies away with her needle; or, if she is writing, it tries to carry off her quill. Sometimes the lady puts a seed or two into a small ivory box, and laying the top lightly on, the bullfinch darts towards it, and dexterously turning the lid upside down on the table, hovers over the open box, from which he picks out the seed, and flies off without alighting.

Another bullfinch (we know not if still in existence,) was purchased some years ago by a lady from a French prisoner. The poor exile had painted the cage of his little captive like a prison, and the bird drew up two little buckets suspended by a gilt-chain, one containing seed, and the other water. This bullfinch was extremely tame, and, though bred in the woods of Greenlaw, (near Edinburgh,) it whistled a variety of troubadour songs.

Bullfinches are affectionate, and get much attached to those who feed them; also to persons they are accustomed to see, if they pay them any attention. We had one which we kept during five or six years. It was mild, gentle, and quiet. A strong attachment took place between it and a green linnet, and this mutual friendship was only broken by death. In a wild state, they frequent woods

and shrubberies, near orchards and gardens, where
they build; sometimes in a beach, or holly hedge,
but oftener in apple or pear trees. These birds
are not migratory, nor are they ever seen in flocks
of more than the parents and their brood. It is
said they are very destructive to gardens, by de-
stroying the young fruit blossoms. This they
probably do in common with many of the hard-
billed birds; but, if we are not misinformed, gardeners blame some of the soft-billed species also
for this mal-practice,—in which allegation we think
gardeners are mistaken. The soft-billed species only
touch those buds that have within them the larvæ
or eggs of insects, and, for this good turn, instead
of being protected, they are often molested and shot.
Bullfinches build their nests early in May, and
have sometimes two broods in the year.

Of the Nest and Eggs.

The nest is composed of dry twigs and a little
wool and moss, lined with the finest fibrous parts
of roots of trees and shrubs, and a few horse hairs.
The eggs, four or five in number, are of a greyish-
white colour, passing into purplish-white, spotted
and streaked with pale blackish-purple and hair-
brown marks.

Of the Young.

Young Bullfinches are rather delicate birds to
bring up. They ought to be ten or twelve days old
before taken from the nest, and then kept dry and
clean, and rather warm. They must be fed early
in the morning, and regularly every two hours
during the day. They will thrive on the same food
as that given to young goldfinches and grey linnets; and old bullfinches are fed with the same
food and managed in the same manner as old goldfinches, &c.

Description and Plumage.

Length nearly six inches; bill, black, short, thick,
and much hooked; eyes, umber-brown; crown of
the head, as far as the eyes, rich glossy velvet-black,
with a tinge of Scotch blue; the black is sometimes
continued from the head under, and round the
bill; neck, and upper part of the back, pearl-grey;
cheeks, lower parts of the throat, and breast, beautiful pale scarlet-red, passing into pale aurora-red,—
the red extending pretty far down below the breast,
terminating in white towards the tail; wings and
tail bluish-black; upper wing-coverts yellowish-

white; tail-coverts, snow-white; legs and feet,
clove-brown. The female has no red on the
breast: it is of a very pale tint, between wood-
brown and hair-brown; and the black on her
head, wings, and tail, is neither so deep, nor the
white parts of the plumage so bright. The young
male bullfinches, for nearly two months, resemble
the female in plumage, and there is no knowing
the male birds from the female till the red appears
on their breast. Too much hemp-seed changes the
plumage of this bird to black: We had one that
was all black except the breast, and even that
had a blackish tinge. Bullfinches have sometimes
been seen white.

Song.

Its common song is a short, often repeated,
agreeable sprightly warble; its call-note is rather
wild, but its soft notes are very melodious. Early
in the morning, or when undisturbed, the bullfinch pours out a soft sweet warbling strain of (at
least to our ears,) charming melody; but so low
is this tender carol, that it cannot be heard at
any distance from the bird: It has also a melan-
choly wailing note which it utters when agitated

by fear. These are its natural notes; but, when young, and properly taught, it will readily learn the sweet, simple national airs of any country. Its ear is good, and its memory excellent, which, combined with its beautiful plumage and sweet natural notes, render this charming bird highly deserving of notice; and, when taught to pipe and whistle favourite national airs, the bullfinch is indeed worthy of a place in the collection of any amateur of song-birds.

GREEN LINNET.

Published by John Anderson Jun.ʳ 55. North Bridge Street. Edinburgh. 1823.

'The woodland choir, sweet denizens of air,
Welcome the sun's approach with songs of joy,
Or, lightly flitting, to the woods repair;
'Or to their mates, on anxious pinions, fly!

But Dick, poor bird! a captive all his life,
The bliss of liberty did never prove;
Powerful, and brave, a stranger yet to strife,
He lived for friendship, and expired.for love!

Anonymous.

GREEN LINNET.

GREEN-FINCH.——GREEN GROSSBEAK.

LOXIA CHLORIS; LINNÆUS.——LE VERDIER; BUFFON.

THIS is a very docile bird. It soon becomes tame, is easily taught a number of very pretty tricks, and grows very affectionate to those who pay it any attention; it is also capable of friendship towards other birds; an instance of which came under our own observation. One we possessed, took a strong attachment to a bullfinch, and the bullfinch returned it. They hung in opposite sides of the same window for a long period; but their attachment commenced shortly after their cages were placed together. We had, besides those two friends, some canaries, mule-birds, grey linnets, and goldfinches.

This green linnet was really a wonderful little creature. Birds often acquire astonishing tricks; but they perform them in a mechanical manner like automatons: Our green linnet, however, in all his actions, seemed to be regulated by something approaching to reason. We used sometimes to let these birds out of their cages all together in a room at one time. The green linnet paid no marked attention to any of them, except the bullfinch, unless when the birds were fighting; and even then merely to put an end to the quarrel, and punish the aggressor. It was a strong, powerful, and spirited bird; but no bully,—not in the least quarrelsome. It submitted to the petty insults of the mule-birds, &c. with great magnanimity and temper, —affording a beautiful lesson to hot-headed, silly-minded man. A strong understanding tends to curb every vice; while ungovernable passion and quarrelsomeness, &c. always spring from some weakness in the mind. Some vice that puts on the semblance of a virtue; such as false honour, for true, manly feeling,—or false courage, (which is often the desperation of a coward,) for true bravery. This last never insults, never offends, is loath to take offence, and is never exerted but in a good cause; whereas those vices are the component parts of that contemptible thing which men of true honour, and true courage, stigmatize, and that justly, with the name of bully.

But to return: The green-linnet paid little attention to the other birds; all his care seemed centered in his friend the bullfinch. " Green Dick," and " Davie," the names of these friends, (and they knew their names, by always coming when called,) continually singled out each other. Green Dick would follow Davie all round the room; if Davie stopped, Green Dick made a peck, and sometimes a jerk, as if to make him move onward, or to start him. If Green Dick succeeded, he appeared delighted,—play evidently being his object. If Green Dick was feeding in any of the cages, and Davie was attacked by any of the other birds, Green Dick instantly flew to his friend, and drove off the assailants. However, we remarked, he never used his bill, but pushed the aggressors away with his breast, as if afraid of hurting them. Green

K

Dick knew each individual of the family in the house quite well; but was attached to one in particular. He would go to her in preference to any one else. If she went into the room where his cage hung, he welcomed her by bunching up his feathers in a particular manner, and by a cheerful note, which he used to none of the other members of the family. If she advanced to the cage, closed her hands, and raised or pushed them forwards as if to hit him, he showed no alarm, but sprung to meet her, and appeared quite pleased; if she put her finger between the wires, and drew it quickly back, as if frightened, he was highly delighted; but if she allowed her finger to remain for him to peck at, he drew his feathers close to his body and seemed greatly disappointed. If she went into the room in a hurry, to fetch any thing without paying the accustomed attention to him, he took the pet, drew himself up, and clapped his feathers close to his body, and, though she went to him in the hope of reconciliation before leaving the apartment, he would take no notice of her. But a little after, if she returned, the coldness was all forgotten, and he welcomed her as before.

At the end of nearly five years, Davie the bullfinch died, during the night. The next morning

Green Dick missed him, (for, as we have already remarked,) the cages hung opposite to each
other at the same window. He uttered a wailing note all that day; the following he was apparently unwell; he put his head under his wing,
and eat no food. At the same time his favourite
in the family was ill: He was brought to her bed
—knew her voice—drew his head from under his
little wing, but his eyes were dim and heavy. He
was carried back to the window where his cage
hung, and the next morning was found dead; this
affectionate bird, to all appearance, having died of
a broken heart.

Green-linnets during winter fly in large flocks,
but in spring they separate: At that time the
males may be seen on the branches of trees, with
their heads always turned towards the sun,—quiv-
ering their wings and pouring out their call-notes.
In spring and autumn they frequent orchards,
gardens, and shrubberies near houses, and are often seen on the high-way, both during winter and
summer, seeking for food. Their food consists of
seeds and grain of different kinds. They make
their nests commonly in hedges or among thick
bushes; seldom far from the habitation of man, often near farm-yards, perhaps (such is the power of

instinct,) for the purpose of being near food when their young are able to fly. The nest is composed of bent, small pieces of dried twigs, and a little moss, lined with hair. The eggs, four or five in number, are of a reddish-white colour, slightly spotted with reddish-orange; somewhat like those of the grey-linnet, but larger. The young may be taken at seven or eight days old. This being a hardy bird, it is easily brought up; and ought to be treated, both young and old, in the same manner as the goldfinch, grey-linnet, &c.

Description and Plumage.

Length nearly six inches,—bill reddish-white, tinged with very pale wood-brown,—eyes umber-brown,—head, neck, back, and wing-coverts a tint between oil-green and siskin-green,—the lower part of the back passing into sulphur-yellow towards the tail,—primary quills, and tail, pale greenish-black, margined with gamboge-yellow,—sides of the neck, and upper parts of wing-coverts, pearl and bluish-grey,—throat, breast, and under parts, siskin-green, inclining to sulphur-yellow,—legs and feet pale flesh-red.

Song.

The common song of this bird is rather loud and monotonous, but not unpleasant. The call-note, which it uses in spring, it also chaunts even in captivity; it is a very loud, whirring, lengthened note. It has also a wailing note, and another very expressive of fear, when any person approaches its nest. But early in the morning, or, when every thing (in nature) is still, this is the time to hear the green linnet's song to advantage; then it has a low, soft, inward warble, as if singing to itself *sotto voce*, and seeming afraid lest its melodious notes should be heard.

YELLOW BUNTING.

YELLOW HAMMER.

EMBERIZA CITRINELLA; LINNÆUS.—LE BRUANT;
BUFFON.

This well-known bird, though rather showy
in its plumage, and very inoffensive in its habits,
meets a foe in every school-boy. Persecuted wher-
ever it builds, no sooner has the yellow bunting
selected a place for its habitation, however artfully
concealed by grass and leaves, than alarm and in-
vasion await it and its brood,—to the latter especially, who seem to have inspired so universal a con-
tempt in the young lords of creation, that a yellowhammer, or as it is called in Scotland, a "*yellow
yite,*" is a common epithet of abuse for a person
who inspires them with ridicule or dislike. This
species haunts lanes and hedges, and frequently

R. Scott Sculpt.

YELLOW HAMMER.

perches on the topmost bough of hedge-rows, ut-
tering its monotonous and simple note; or flitting
along in short interrupted flights, as if incapable of
escaping from the passer-by. It generally builds
in May, and makes choice of a low bush or hedge
for that purpose, though we have seen a nest in a
moist mossy bank above a streamlet, canopied by
a plant of avens, the decayed leaves of which laid
the foundation for the nest, while the green foliage
and bending flowerets concealed the artless dwelling from public view. But the yellow bunting
rarely builds on the ground, preferring a low bush,
or among reeds in moist places; " where grows
the willow and osier dank." In winter it assem-
bles in flocks with the chaffinch, the sparrow, &c.
to pick up grain, rape-seed, &c. in stack-yards.

Few, we believe, keep the yellow hammer as a
bird of song. Some years ago we possessed one
which we reared from the nest; it was extremely
docile, and showed great aptness in learning what-
ever we attempted to teach it.

This was a lively, playful bird even in confinement, except only, when after enjoying restrict-
ed freedom, it returned to its cage; then, for a
few minutes, it appeared sad and listless, but shortly after resumed its vivacity. When it was permit-

ted to fly through the room, where the cage stood, it would (during breakfast,) perch on the sugar basin, nibble at the sugar, and peep into every cup. When any of the females of the family were at work, it delighted to hover about their fingers, and try to seize the thread, by pulling it away; if it met with no resistance, it soon ceased from its labours, or, if it succeeded in carrying off the prize, the pleasure was at an end,—for, dropping the thread, it soon returned to the sport with renewed interest, evidently more intent on pastime than plunder. We often placed the open cage outside the window, when the bunting would hop out of the cage, perch on the top; and, after looking about it for some time with apparent unconcern, instead of availing itself of the liberty thus offered, it would fly back to the room.

This elegant little bird, in a wild state, is almost as familiar and domestic as the common house sparrow; it frequents hamlets and villages, and never builds far from the habitations of man. It may be often seen on the high-way, hopping and flitting before the pedestrian, seeking food for its mate or its young; it also frequents the high-roads in winter in pursuit of food, and during that season becomes so tame as to allow a person to ap-

proach within a few paces of it, before it flies off,
and then only to a very short distance.

When any one approaches their nest, they utter
a plaintive, wailing note; which, even to man, conveys an expression of the feeling of pain.   The female is very tenacious of her nest, and sits so close
that she may often be taken on the eggs with the
hand.

Description and Plumage.

The length of this elegant species is about six
inches,—bill bluish-grey, slightly tinged with
brown,—eyes umber-brown or hazel,—head rich
gamboge-yellow, streaked with umber and blackish-brown,—throat, breast, and under parts, pale
gamboge-yellow; breast streaked with wine-yellow,
passing into yellowish-brown,—back and wing-coverts honey-yellow, inclining to yellowish-brown,
—the feathers of these parts marked in the centre with umber-brown,—edges of the primary
quills sulphur-yellow; the rest of the quill feathers,
and tail, blackish-brown, the last edged with yellowish-white, and a little forked.

The female is very like the male, only the yellow on her head is not so bright; and indeed her

plumage is altogether more of a dingy hue. The young birds have no yellow in their plumage for a month or two; they resemble the female, but their plumage is rather more dingy.

Song.

The song of the yellow hammer is hardly worth a particular notice; nor should we have assigned it a place among the song-birds, were it not said to be capable of acquiring the notes of other birds. Its own song is not unlike that of the common bunting, but more sweet, though it consists of only a few notes often repeated.

The chirp or cry of the young is very like the notes of the young of the hedge sparrow or red-breast.

THE BLACK-HEADED BUNTING,

OR REED-BUNTING.

EMBERIZA SCHŒNICLUS; LINNÆUS.—L'ORTO-
LAN DES ROSEAUX; BUFFON.

THE favourite haunts of this beautiful bird are
fens, bogs, and marshes, where, among reeds and
bulrushes, it forms its nest. It was formerly said
to be a fine song-bird; but modern writers allege
this assertion to have originated from its being
mistaken for the sedge-warbler: And Montagu
declares, "that the reed-bunting has no song
worthy of notice, but that it consists of two sharp
notes three or four times repeated, which it continues for a length of time, and, when alarmed by
a sportsman, or a dog, it utters a harsh incessant

cry, which is so discordant as to frighten away game and other birds; but we know that the song of the black-headed or reed-bunting is very superior to that of any other British species of bunting, the snow-flakes perhaps excepted; but this last we never heard.

The black-headed bunting is a timid bird, and so easily alarmed that it is difficult to take it; and indeed the ground which it frequents, and where it breeds, is not very easy of access: it is impossible to walk with " a stealthy pace" or silent footing in a marsh:—but, like most of the other species of buntings, it occasionally resorts to corn-fields and roots of hedges, in pursuit of food, repairing to the fens, bogs, and marshy places in the evening; and, during long-continued rains, it sometimes betakes itself to the high grounds, and perches among the bushes.

The black-headed bunting is not a gregarious bird; indeed, more than three or four have never been seen together, or at most a brood-flock. On the continent it is a migratory bird, arriving there in spring and departing in autumn; but in Britain it remains the whole year.

Of the Nest, Eggs, &c.

The nest of this bird is most frequently situated near swampy places, though occasionally it may be fixed in a low bush of willow, or even furze, a foot or two from the ground, at a short distance from any marsh or fen. The nest, however, is generally placed amongst clumps or bunches of long grass, willow-roots, tufts of rushes, reeds, &c. It is a flimsy structure, composed of bent and withered grass, and slightly lined with a few horse hairs. It contains four or five eggs of a dingy bluish-white, mottled, and streaked with umber-brown, inclining to blackish-purple, in resemblance somewhat between those of the yellow bunting and those of the chaffinch.

The black-headed bunting is somewhat less than the yellow hammer; length about six inches; bill, bluish-grey; eyes, umber-brown or hazel; the head, round the eyes, throat, and part of the breast, velvet-black. From the bill on each side to near the back part of the neck, runs a band of yellowish-white; neck, back, wings, and tail, yellowish-brown, each feather streaked down the middle with dark umber-brown; primary-quills edged with very pale wood-brown; outer edge of the outer feathers of the tail, snow-white; legs and feet dull yellowish-brown.

Before quitting the buntings, we may here mention the snow-flake, snow-bunting, or *emberiza nivalis*, as it is said by some authors to sing delightfully in its summer haunts, which are in high northern latitudes, as Spitzbergen, Lapland, Hudson's Bay, &c. It visits us only in winter, and returns northward in spring to breed; so that its song is never heard in this country. But, as it is said to be such a charming songster, it would be worth while to try the experiment. They could be easily caught, as they frequent the northern parts of our island, during winter, in vast flocks. This bird never perches, but runs on the ground like a lark, and, like that bird, its hind claw is very long. The nest is placed in the crevices of rocks, and formed of dried grass, lined with feathers and the down of the arctick fox. The eggs, four or five in number, are bluish-white, spotted with umber-brown. Bewick says, these birds "sing sweetly, sitting on the ground." " On their first arrival in this country they are very lean, but soon grow fat. The Highlands of

Scotland abound with them." They are also found in Northumberland. Their plumage is beautifully pied with snow-white and velvet-black. It is nearly seven inches in length, elegantly shaped, and altogether a very pretty bird. The canary-bunting, which has been but rarely found in Britain in summer, is said by some authors to be the snow-flake in its summer plumage.

FINIS.

A. Allardice, Printer, Edinburgh.

9 781014 728739